CET Exam Book

3rd Edition

CET Exam Book

3rd Edition

Dick Glass
Ron Crow

TAB Books
Division of McGraw-Hill, Inc.
New York San Francisco Washington, D.C. Auckland Bogotá
Caracas Lisbon London Madrid Mexico City Milan
Montreal New Delhi San Juan Singapore
Sydney Tokyo Toronto

pbk 4 5 6 7 8 9 10 11 12 13 DOH/DOH 9 9 8 7 6 5 4
hc 3 4 5 6 7 8 9 10 11 12 DOH/DOH 9 9 8 7 6 5 4 3

Library of Congress Cataloging-in-Publication Data

Glass, Dick.
CET exam book / by Dick Glass and Ron Crow. — 3rd ed.
p. cm.
Crow's name appears first on the earlier edition.
Includes index.
ISBN 0-8306-4069-X (hard) ISBN 0-8306-4068-1 (pbk.)
1. Electronics—Examinations, questions, etc. 2. Electronic technicians—Certification—United States. I. Crow, Ron.
II. Title.
TK7863.G55 1992
621.381'076—dc20 92-13195
CIP

Acquisitions Editor: Roland Phelps
Editor: Melanie Brewer
Director of Production: Katherine G. Brown
Book Design: Jaclyn J. Boone
Cover Design: Denny Bond, East Petersburg, Pa.

EL1
4199

Contents

Introduction

IN THIS THIRD EDITION OF THE CET EXAM BOOK THE AUTHORS HAVE enlarged the material pertaining to the Associate examination. In addition, each of the Journeyman Option examinations has an entire chapter devoted to it.

The writing of this guide comes immediately following publication of all-new CET exams by the Electronics Technicians Association, Int'l, Inc. (ETA-I).

Chapter one outlines the details of the CET program. It explains how you can qualify. A membership application is also included because many professionals recognize the importance of belonging to their professional society. You are invited to become a member, even prior to becoming certified.

Also listed are locations where CET exams are frequently held. Unpaid Certification Administrators (CAs) are available for you to set up a time to attempt the exam. The success of the CET program is primarily due to the dedication of these test monitors. When you feel ready to attempt the exam, contact the CA nearest you. If none are nearby, you do not have to travel hundreds of miles for the exam. Call ETA's headquarters (1-317-653-8262). ETA's staff will work with you to locate an electronics school near you to administer your examination.

American military personnel can contact their education office. Through the DANTES (Defense Activity for Non Traditional Educational Services) program, the CET exams are ordered and a test session planned for you on your base or ship.

You will find a quiz in most chapters. When you can pass these and feel you really understand the subject matter, you are ready to take the examination and become a CET. Good luck.

1
CET history

SINCE 1966 THE CERTIFIED ELECTRONICS TECHNICIAN PROGRAM HAS continued to grow in importance. The founders originally were attempting to provide an examination and experience criteria that matched those required in the U.S. Labor Dept. (Bureau of Apprenticeship and Training) four-year, skilled-trades projects of that time.

NEA (National Electronic Associations) was the sponsoring organization. Its membership was composed primarily of television shopowners and technicians. Four years of experience and/or electronics schooling was established as the time requirement. The examination consisted of multiple-choice questions pertaining to antenna theory and wave propagation; basic electricity; electron tubes and solid-state devices; electronics fundamentals; color codes; safety; common circuits; radio and TV block diagrams, circuitry, operation, and servicing.

While similar examination and registration programs for electronics technicians were being instituted by regional technician or dealer groups in Washington state and California, Edward T. Carroll, CET, was awarded the first CET certificate (IND-1) under the NEA program, in 1966. Shortly thereafter the California and Washington groups adopted the national program. The Oregon State Professional Licensing Agency adopted the CET exam for its state licensing program. Some manufacturing companies and a number of commercial and public technical colleges joined in the effort to help NEA, and later ISCET (International Society of CETs) to expand the CET effort. The CET program was never funded (as was the CAE project in the auto industry). The CET program has always been financed by test fees only, through NEA, then ISCET and ETA-I.

Soon CET was accepted as a valuable education and recognition project. It appeared to benefit technicians, service business owners, manufacturers, other suppliers, and the educational community. The need for study materials aimed at helping technicians bring themselves up to the CET level was apparent. CET study guides were an immediate success in the book world.

In 1978 a portion of the CETs in ISCET decided to form ETA. It was incorporated as a nonprofit professional association under the laws of Indiana. Those early organizers believed a professional association, rather than a business league, would be more attuned to technician needs. ETA immediately set up its own certification program, with Ron Crow, CETma, as director. The ETA offices were located at Iowa State University in Ames, Iowa until 1990, when the program shifted to ETA headquarters in Greencastle, Indiana.

During the period of 1966 to 1978, Leon Howland, CET, was director of the original CET program, working out of NEA's headquarters on South New Jersey Street in Indianapolis. For a time Howard Bonar, CET, of Muscatine, Iowa, was the director. Later, ICS (International Correspondence Schools) of Scranton, PA, performed grading and accrediting. Then, after NEA evolved into NESDA, grading and administration took place in NESDA's Indianapolis headquarters under the direction of Executive Vice President Dick Glass, CETsr (see Fig. 1-1).

1-1 CSI certificate presentation.

During these years electronics grew at a phenomenal rate. Technicians began to specialize. Industrial electricians were joined by industrial electronics technicians. Specialties in medical, computer, communications, and other areas emerged. To bring some order out of the complex evolution of electronics, ETA continued the Associate level certification for technicians who had not yet acquired four years of experience and/or schooling. ETA then added options in six categories. It soon became apparent that the technology as well as the terminology used to describe the work of electronics technicians was not going to settle down; instead it appeared to be a gradually changing profession. Therefore, the following categories for CETs were established:

CET choices

Journeyman At least four years experience and/or schooling. Passing score of at least 75 percent on written examination.

Senior Eight years experience and an 85 percent passing grade on examination.

Master Four or more years experience and/or schooling and a 75 percent score on examinations in at least six option categories.

Associate Less than four years experience and/or schooling, and a score of at least 75 percent on the Associate Level examination.

Journeyman options

Consumer Electronics TV—VCR—Audio—AM/FM Radio.

Computer Computers and digital electronics.

Industrial Environmental and Process Control—Motorized equipment.

Bio Medical Medical Electronics hardware.

Radio Communications Transmitters—Broadcast principles—Business and two-way Radio-Antennas.

Telecommunications Telephone—Data—Cellular—Microwave communications.

Video Distribution Master Antenna systems—MATV—Satellite—Cable TV.

Satellite Home and Commercial Satellite reception equipment.

Endorsement to radio communications

Avionics This is not an option, but an endorsement dealing with air to ground, air to air, and navigation equipment.

Other certifications

CSI Certified Satellite Installer (for professionals engaged in satellite installation and servicing).

CSS Customer Service Specialist. (Certifying knowledge of customer relations and handling for electronics technicians. Requires acceptance of a Code of Conduct.)

ETA certification program procedures

Journeyman CET

The Journeyman level CET is the status originally intended for all CETs in 1966. It remains the category with the highest number of CETs. Those who attain this level use the CET designation following their name: John Jones, CET. The journeyman exam consists of two separate parts: The first is the associate section. It is made up of testing materials every tech should know. This includes 75 questions about basic electronics, transmission cable, common circuits, component recognition, basic digital circuitry and components, test equipment usage, color codes, ac and dc theory, antennas, and wave propagation.

The second portion, consisting of between 25 and 35 questions, is the option section. You can choose which option you wish to take; if you are working in computer servicing, you most likely would choose the computer option. Combined, you should expect 100 to 110 multiple-choice questions. With a passing score of 75 percent or more, along with meeting the time and experience requirements, you are entitled to the CET rating. You also receive a wall certificate and CET wallet card.

Senior CET

Because electronics has evolved at an ever-increasing rate, technicians need to continually enlarge their knowledge of the technology and servicing procedures needed to accomplish their jobs. Being a "crackerjack" technician in all major areas of electronics is extremely difficult. Many CETs realized that they must keep learning to stay competent in their present work. Few other crafts or professions in history have made such demands on their workers.

To provide a continuing credibility to the CET program, the Senior CET level was created in 1972. Certified technicians might elect to retake the CET exam again, after they have acquired eight years of experience or schooling. With a slightly higher test score (85 percent, the technician can qualify as a Senior CET: John Jones, CETsr.

While the original intent of the Senior-level program was to allow recertification of previously certified technicians, many first-time CETs believed they should qualify as seniors if they made the required 85 percent grade and could show at least eight years of experience or schooling. This addition was accepted by both the CET and ETA programs' directors and is allowed.

Master CET

Many electronics instructors and some technicians, especially those who have job experience in more than one area of electronics, are qualified to be certified in more than one category. Many techs do maintain certifications in more than one option. A few can pass the option exam in all the categories. The 75 percent score on every option was originally required in ETA. Now, a 75 percent score on at least six options is required to become a Master CET: John Jones, CETma.

Associate CET

The Associate exam was established to allow graduating electronics school technicians and others with less than four years of experience or schooling an opportunity to prove their knowledge. Technical school graduates should be able to pass the Associate exam. If successful, the ACET designation is granted and the technician is listed in the ETA Journal, receives an Associate wall certificate and a wallet card.

A technician cannot take a Journeyman option without at least four years of experience or schooling; if the technician has less than four years of experience, only the Associate exam can be attempted. If the tech has acquired four or more years of experience, then the Associate plus an option of his/her choice, must be taken (or the Senior or Master option).

Exceptions

Occasionally, a tech might feel fully capable of passing the Master or Senior exam on the first attempt. In this example, if he fails to achieve 85 percent, as a Senior, (but does make at least 75 percent), or he might pass one or more Master options, but fail others, then the examinee can request a journeyman certificate for a slightly higher fee.

Review

ETA has a computerized program to provide review of missed questions on the CET exams. The readout returned to the examinee, for an additional fee, does not relate the correct answer to any question, but it does offer suggestions as to why the answer given was wrong. The review program was installed in response to requests by examinees and test proctors over the years to help point out weak areas a test take might have.

2
Basic mathematics

ELECTRONICS IS AN "INHUMAN" SCIENCE; IT IS DIFFICULT TO LEARN. Because of this we try to relate electronics to things we do understand. For example we compare electric voltage to the water pressure in a garden hose. We try to understand oscillators by comparing them with a child's swing.

Many of the concepts in electronics can only be understood by using arithmetic, algebraic formulas, ratios, and just plain addition, subtraction, division, and multiplication. A technician must accept this and let math become his/her friend; that's the bad news.

The good news is that much of the math used in learning electronics is rarely needed again in the technician's career. Most of an electronics technician's work is hands on. By working problems out mathematically, you learn the relationships you must know to get the feel of electronics circuitry. Math is used on each page of the CET exam. Included below are ten questions covering much of that which you should become comfortable with; you will encounter these concepts throughout your career. Another quiz is included in this chapter following the explanation for these ten.

Quiz

Q1. The most basic electronics formula is:

$$I = \frac{E}{R}$$

If $E = 40$ volts and $R = 10$ ohms, what is I?

a. 4 amps
b. 4 milliamps

c. 40 amps

Q2. $I = 0.01$ amp, $R = 1000$ ohms. What is the value of E?

a. 10 volts
b. 0.00001 volt
c. 100 volts

Q3. One of the power formulas is: $P = EI$. If $R = 20$ and $I = 0.001$ amp, what is the value of P?

a. 0.0002 watt
b. 0.002 watt
c. 0.00002 watt

Q4. The formula for inductive reactance is: $X_L = 2\pi fL$. If the frequency (f) goes up, what happens to the inductive reactance?

a. It goes up.
b. It goes down.
c. It does not change.

Q5. $X_C = \dfrac{1}{2\pi fC}$

$C = 0.01$ farad; $f = 1000$ Hz. What is X_C?

a. 159.2 ohms
b. 0.01592 ohm
c. 15,923 ohms

Q6. The formula for wavelength in feet is:

$$\lambda = \frac{984{,}000}{f\ (\text{kHz})}$$

What is the wavelength of a 1500-kHz radio wave?

a. 0.656 foot
b. 6560 feet
c. 656 feet

Q7. Parallel impedance totals are calculated using the formula:

$$Z = \frac{Z_1 Z_2}{Z_1 + Z_2}$$

If $Z_1 = 684$ and $Z_2 = 1000$ ohms, what is the total impedance?

a. 406 ohms
b. 4060 ohms
c. 40.6 ohms

Q8. A capacitor charged to 100 volts discharges in 14 milliseconds to 37 volts through a 4700-ohm resistor. Because $T = RC$ (That's the RC-time formula), the capacitance is:

a. 336 farads
b. 2.98 microfarads
c. 5000 microfarads

Q9. The formula for resonance is:

$$f = \frac{1}{2\pi\sqrt{LC}}$$

If $L = 0.01$ henry and $C = 1.0$ microfarads, what is f_r?

a. 15.924 MHz
b. 1590 Hz
c. 31 MHz

Q10. A meter movement to be used for 5 volts full-scale deflection requires 50 microamperes for full-scale deflection. The internal resistance of the movement is 5 kilohms. What must be added to the meter movement to measure 5 volts?

a. A 95-kilohm resistance in parallel.
b. A 105-kilohm resistance in parallel.
c. A 95-kilohm resistance in series.
d. A 105-kilohm resistance in series.

Quiz explanation

If you did well on this portion of the exam you should find additional math no more difficult. If you had difficulty on these questions, go back and work on those that bothered you until that type of math starts to come easy for you. If you have trouble transposing a formula, work out some similar problems until you find them easy and until you know they are correct. Part of the problem with electronic math is that you are dealing in extremely small amounts, such as 0.01 microfarad, and in extremely large amounts, such as 2.8 gigahertz. Not only do you need to know how to work math equations, but you also need to be able to convert the maxi and mini quantities to figures that are easier to handle in electronics formulas, starting with Ohm's law. If you make up your mind to master the math so that it doesn't bother you on the exam, then your only problem will be to recall the applicable formula or relationship of the electronic components.

The mathematical relationships in electronics are seemingly endless. To try to remember them all is impossible. The trick is to sort out those that are commonly used and not to cloud your mind with those that should be obtained from a reference book when they are needed. This book focuses on those concepts that are commonly needed. Be sure you firmly implant them in your mind prior to the exam.

Question 1 gives you the most basic formula technicians use.

$$I = \frac{E}{R}$$

All you need to do is divide E (40) by R (10) to get the answer: 4. Unlike many electronics formulas, this one is straightforward: I is in amps; E in volts; and R in ohms.

Question 2 is slightly tougher; it deals with decimals rather than integer numbers. $E = 0.01 \times 1000$ ($0.01 \times 1000 = 10$). Therefore $E = 10$.

Question 3 seems simple. The only problem is that the $P = EI$ formula has no R in it; instead it has an E. You must remember that P also $= I^2R$. You also could change the $P = EI$ formula to $P = IR \times I$ (remember that $E = IR$, as in Q2) or, $P = I^2R$. I^2 is $I \times I$, or 0.001×0.001, or 0.000001). Thus $0.000001 \times R$ (20) equals 0.00002, or answer c. You might prefer to work math problems using powers of 10. In that case, multiply (0.001) or $1 \times 10^{-3} \times 1 \times 10^{-3}$, which gives you $1 \times 10^{-6} \times 2 \times 10^{+1} = 2 \times 10^{-5}$, or 0.00002.

Does question 4 seem too simple? Some technicians have trouble with this relationship. Anything on the right side of the formula that increases has to increase the value of X_L. If X_L is truly equal to the right side of the formula, then if the right side gets bigger, the left side must likewise increase. It is easy to confuse this formula with its sister formula—the formula for capacitive reactance

$$X = \frac{1}{2\pi fC}$$

Also, many are intimidated by pi (π). Remember π is simply a quantity: 3.14, and 2π is 6.28. If you really want to know why pi is 3.14, check out the formula for the area of a circle: $A = \pi \times$ radius squared. There are 360 degrees in a circle. There are 360 degrees in the swing of an ac sine wave. Study geometry, because the current and voltage in capacitors and inductors use some of the same relationships. The quantity π = a constant: 3.14. The formula in Q4 is easy; X_L goes up when frequency goes up. The confusing thing about this simple formula is that you might get it tangled up with inductance, or L. Inductive reactance is not inductance. Keep the two separated in your mind, and you won't have trouble with problems of this kind.

Question 5 is the opposite of the previous question and formula. Capacitive reactance gets smaller if f or C gets larger. ($1/10$ is smaller than $1/2$ isn't it?) The problem here is to find $2\pi fC$ so that we can get rid of that awful fraction. We do that by dividing $2\pi fC$ into 1. Use your calculator; they are allowed. Multiply 6.28 (2π) times 0.01, times 1000. The answer is 62.80. Divide 1 by 62.80. The answer is 0.01592 ohm. $X_C = 0.01592$ ohm. You couldn't merely guess at that answer. (The answer of 0.01592 ohm is not much resistance anyway.)

While the big numbers in Q6 concerning wavelength are difficult, the real problem is that few of us remember the formula for wavelength. Even if you know the formula, you can't easily know that our forefathers decided that in this formula, f, or frequency, should be in kilohertz! So you can easily get the wrong answer even if the math is right. Around the world, a more common formula for wavelength is $300{,}000/f$ (kHz). This gives you the same answer, but in meters rather than feet. There is no easy way to remember this concept, just put it in a separate brain cell and never forget it.

The formula in Q7 can be used to figure impedances or resistances in parallel. If you have more than two parallel paths, you can calculate two of them, then use the resulting answer to work out the next, or third leg in combination with one and two, and so forth.

The concepts in Q8 concern the time constant formula: RC time. You can't figure it out; that formula must be memorized. $T = RC$ is not the most difficult formula you will have to remember. To make it tougher, this question asks what C is, so we not only need to know the formula, but also how to transpose the equation. $T = RC$ means RC time equals R in ohms times C in farads.

The next thing we need to know about RC time is that a capacitor will charge up to 63% of its full value (or discharge up to 37% from its full value) in just 1 RC time. It takes 5 RC times for it to charge to nearly 100% (actually 93.3%). Because Q8 tells us it takes 14 milliseconds to discharge from 100 volts to 37 volts, one RC time is 14 milliseconds. If you have all the above now firmly implanted in your memory cells, then all you have to do is work the math to get the correct answer. $T = RC$; $0.014 = 4700 \times C$; $0.014 = 4700C$; $C = 0.014$ divided by 4700; $C = 0.000002978$, or 2.98 microfarads.

The only thing tough about Q9, because the formula is given, is that you end up with 0.00000001 under the radical. The rest is merely math.

The first thing to realize in Q10 is that you need a resistance in series with the meter movement. If you try a parallel resistance, you will see that applying 5 volts will result in 5 volts divided by 5 kilohms (meter resistance) amps, or 1 milliampere through the 50-microamp meter movement! That would cause damage to the D'Arsonval meter movement coil. With a series circuit required then, the current through the circuit at the applied 5 volts must be the 50 microamps needed for full-scale deflection. The total resistance needed is: 5 volts divided by 50 microamps, or 100 kilohms. This leads to the series resistance being 100 kilohms minus 5 kilohms (the resistance of the meter movement itself), or 95 kilohms.

Additional math questions

Now that you have mastered the concepts in the previous ten questions, try the following 19 quiz samples. These are more complex and a dozen of them require you to understand digital math. If you can master these, you will have no trouble with any math questions on the CET exam.

Q1. When the frequency is increased in an inductive circuit the inductive reactance:

a. increases
b. decreases
c. remains the same

Q2. In the circuit shown in Fig. 2-1, R equals:

a. 20 kilohms
b. 5 kilohms
c. 5 ohms
d. 20 megohms

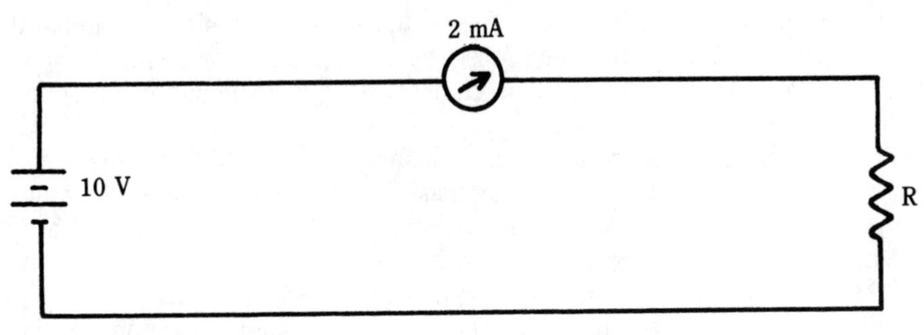

2-1 Circuit for Ohms Law problem in Q2.

Q3. In the circuit shown in Fig. 2-2, V_{R2} (voltage across R2) equals:

a. 10 volts c. 5 volts
b. 6.67 volts d. 4 volts

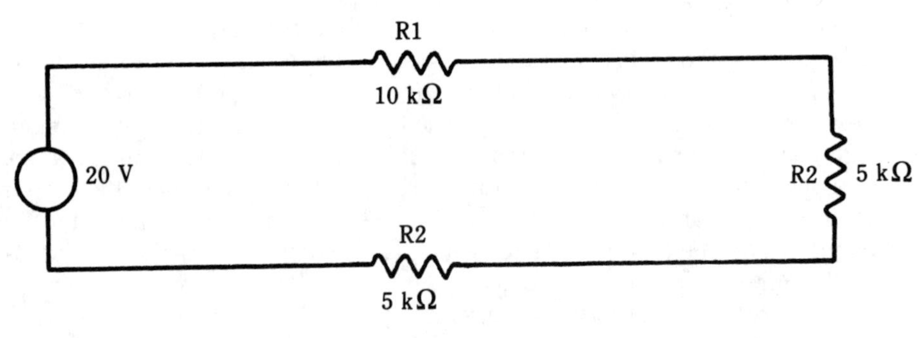

2-2 Series resistor circuit for Q3.

Q4. $C = \frac{KA}{d}$

As this formula explains, the capacitance of a parallel plate capacitor is proportional to the area of the facing plate and the dielectric coefficient K, and inversely proportional to the distance, d, between plates. Solving this formula for d gives d equal to:

a. C/KA c. CKA
b. CK/A d. None of these.

Q5. In a certain transmission line the wave front travels at a velocity, v, 65% of the speed of light. If the speed of light is 186×10^3 miles/second and a 50-MHz signal is sent down the line, one wavelength, λ will be:

$$\lambda = \%\left[5280\left(\frac{V}{f}\right)\right]$$

a. $\dfrac{50\times 10^6}{(186\times 10^3)(0.65)5280}$ feet

b. $\dfrac{(186\times 10^3)(0.65)5280}{50\times 10^6}$ feet

c. $\dfrac{50(5280)}{(186\times 10^3)300}$ feet

d. None of these.

Q6. If in the power formula $P = EI \cos\theta$ for ac circuits, θ can vary from 0 to 90°, $\cos\theta$ varies from:

a. 1 to ∞
b. 1 to 0
c. 0 to ∞
d. None of these.

Q7. If the phase angle in Q6 is 90°, this means that the circuit would have no power dissipation. In other words it would be purely:

a. reactive.
b. resistive.

Q8. The number 1101 in binary is equivalent to a decimal number of:

a. 13
b. 11
c. 25
d. None of these.

Q9. Add the binary numbers: 101101 + 110100:

a. 1100001
b. 1011001
c. 111001
d. None of these.

Q10. In counting numbers in a system with a base (or radix) of 8, the next three numbers after 7 will be:

a. 8, 9, 10
b. 8, 10, 11
c. 10, 11, 12
d. None of these.

Q11. The binary number 110101 is equivalent to an octal number of:

a. 35
b. 65
c. 53
d. None of these.

Q12. The hexadecimal number 2*E* is equivalent to the decimal number of:

a. 46
b. 37
c. 16
d. None of these.

Q13. The hexadecimal number C_4 is equivalent to a binary number of:

a. 11110100
b. 10100010
c. 01110111
d. None of these.

Q14. The Boolean statement, *A* AND *B* equals *Q* can be written:

a. $AB = Q$
b. $A \bullet B = Q$
c. $(A)(B) = Q$
d. All of the above.

Q15. The Boolean statement $A \bullet \overline{A}$ is the same as:

a. 1　　c. A
b. 0　　d. $\overline{A}$

Q16. Reduce the following Boolean expressions to a simple form:

a. $ABC+AB\overline{C}$ = ______
b. $X(X+Y)$ = ______
c. $(G+H)(G+K)$ = ______
d. $\overline{A}\,\overline{C}+\overline{A}C+BC$ = ______
e. $\overline{A}B+\overline{A}\,\overline{B}+A$ = ______
f. $\overline{A}\,\overline{B}(A+B)$ = ______

Q17. If the three inputs to an OR gate are "1", "0", "0" respectively the output will be:

a. 1 volt　　c. "1"
b. 0 volt　　d. "0"

Q18. $\overline{A \bullet B}$ describes:

a. a NOR gate.　　c. an AND gate.
b. a flip-flop.　　d. a NAND gate.

Q19. An exclusive OR can be represented as $A \oplus B=F$, or it can be written as:

a. $A\overline{B}+\overline{A}B=F$　　c. $(A+\overline{B})(\overline{A}+B)=F$
b. $AB+\overline{A}\,\overline{B}=F$　　d. None of these.

Formulas

Several of the quiz questions deal with the manipulation of formulas. Q1 requires that the formula for inductive reactance be recalled and that the relationships be understood in an algebraic equation. In this case $X_L=2\pi fL$ is the formula you should remember. Increasing f will increase the product of $2\pi fL$, hence X_L will increase. The companion formula for X_C is

$$X_C=\frac{1}{2\pi fC}$$

Here if frequency is increased the product $2\pi fC$ again increases; but, because the product is in the denominator the quotient $1/2\pi fC$ decreases.

Question 2 requires some manipulation of Ohm's law. Because we are looking for R, the formula $E=IR$ can be changed to

$$R=\frac{E}{I}$$

by proper algebraic procedures. Substituting $E=10$ volts and $I=2$ milliamperes gives:

$$R = \frac{10\text{ V}}{2\text{ mA}} = \frac{10}{2 \times 10^{-3}} = 5 \times 10^3 = 5\text{ k}\Omega$$

Question 3 is more manipulation of Ohm's law. To find the voltage across R2, first find the series current flow.

$$I = \frac{E}{R_T}$$

where R_T is the total series resistance $R_1 + R_2 + R_3$

$$\begin{aligned} I &= \frac{20}{10\text{ k}\Omega + 5\text{ k}\Omega + 5\text{ k}\Omega} \\ &= \frac{20}{20 \times 10^3} \\ &= 1\text{ mA} \end{aligned}$$

The voltage drop $V_{R2} = IR_2$

$$\begin{aligned} &= (1\text{ mA})(5\text{ k}\Omega) \\ &= 5\text{ V} \end{aligned}$$

You can expect any kind of question dealing with the manipulation of formulas. You might be expected to recall from memory the more important formulas, such as Ohm's law and the reactance formulas.

Question 4 contains a formula and you are expected to manipulate the algebraic terms and solve for one term (*d*). Remember that an equation remains equal if we do the same thing to both sides of the equation:

$$C = K\frac{A}{d}$$

$(d)C = K\frac{A}{d}(d)$ Multiply both sides by *d*.

$\frac{dC}{(C)} = \frac{KA}{(C)}$ Divide both sides by *C*.

$d = \frac{KA}{C}$ Answer d, none of these, is therefore correct.

In Q5 a formula is again given but some conversions and substitutions are necessary. The formula to determine wavelength in a cable is $\lambda = v/f$ where v is the velocity of propagation and f is the frequency. But ($v = 0.65c$) where c is the speed of light ($\approx$ 186,000 miles/sec.)

$$\text{hence } \lambda = \frac{0.65(186{,}000)}{50 \times 10^6}\text{ miles}$$

but to convert to feet, 1 mile = 5280 feet.

$$\lambda = \frac{0.65(186{,}000)}{50 \times 10^6} \times 5280 \text{ feet}$$

$$\text{therefore } \lambda = \frac{186 \times 10^3(0.65)5280}{50 \times 10^6} \text{ feet}$$

Question 6 requires knowledge of the relationships in a right triangle. These are defined as shown in Fig. 2-3.

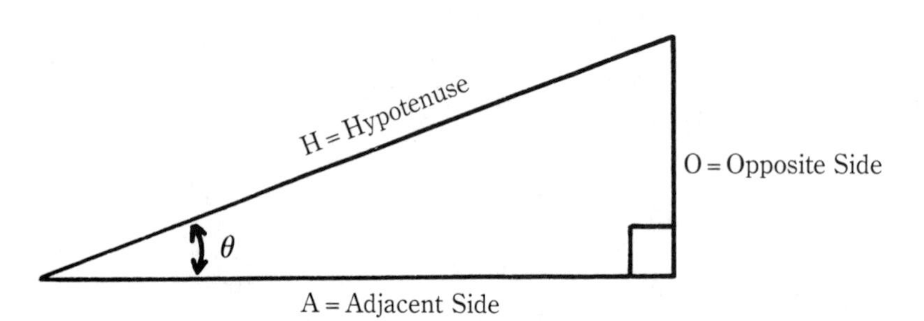

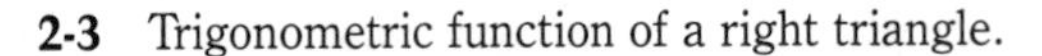
2-3 Trigonometric function of a right triangle.

θ varies from 0° to 90° as stated in Q6. At 0° the opposite side is 0 and $A = H$ hence:

$$\cos \theta = \frac{A}{H} = \frac{A}{A} = 1$$

At 90° the adjacent side equals 0 and $O = H$ hence:

$$\cos \theta = \frac{0}{1}$$

therefore $\cos \theta$ varies from 1 to 0.

Some meanings of the power formula $P = EI \cos \theta$ for the relationships in the formula, are covered in Q7. Because the $\cos \theta$ at 90° is equal to zero, the power dissipated in the circuit must be zero because $EI\,(0) = 0$. The only way for this to occur is to have no resistance in the circuit. Hence the circuit is entirely reactive. To say it another way: if the circuit has any resistances the phase angle θ cannot be 90°. Also notice if the circuit is resistive $\theta = 0$, $\cos \theta = 1$ and the formula reduces to the familiar $P = EI$.

Binary mathematics

Binary mathematics is really no different than mathematics in our more commonly used decimal system. All the rules are the same. The problem is that most of us forgot the rules a long time ago. We are so used to numbers with place values that we give little thought to the structure of our decimal system. In binary math only two digits are available for counting. The symbols for these two digits were chosen as 0 and 1, the same symbols used for two of the digits used in the decimal system. To designate the nothing situation, 0 is used and 1 designates a single unit.

In the decimal system we have ten digits: 0, 1, 2, 3, 4, 5, 6, 7, 8, and 9. These are symbols used to designate these digits and their meanings are familiar to us all. If we desire a quantity of items designated as 5, we all know how many items to expect. If we desire to designate a quantity of items more than 9, we use a system of numbering that gives value to the position of the digits with respect to each other. The digits arranged as "21" means that we have 1 "unit" plus 2 "tens." The number 5941 means we have 1 "unit," 4 "tens," 9 "hundreds," and 5 thousands." We even give the numbers names to help us keep track. This last number, for example, is named five thousand nine hundred forty one. Our naming system takes into account the place value. In the binary system all the same rules apply. Table 2-1 shows a comparison of a few numbers in decimal and binary.

Table 2-1. Counting with only two symbols, 0 and 1.

Binary Number	Decimal Number Equivalent
0	0
1	1
10	2
11	3
100	4
101	5
110	6
111	7
1000	8
1001	9

In binary "place" also has value just as it does in a decimal system. The place value is different however. In decimal the place values are units, tens, hundreds, thousands, and so forth. Each place named, to indicate the value that a digit in that place is multiplied by to obtain the total place value. For example in 3406 the fourth place indicates that there are 3 times 1000 counts in this number. The third place indicates there are 4 times 100 counts in this number. No tens (0×10) are indicated by the second place and 6 units (6×1). In binary the place names are units, twos, fours, eights, sixteens, and so forth with each place value double the last.

Question 8 concerns changing a binary number to a decimal number. The number given is 1101 in binary. In this number there is one unit, no twos, one four, and one eight. $8+4+1=13$. If this is unclear, practice counting in binary. Other examples using base 10 and base 2: $1_{(10)}=1_{(2)}$, $2_{(10)}=10_{(2)}$, $3_{(10)}=11_{(2)}$, $4_{(10)}=100_{(2)}$, $5_{(10)}=101_{(2)}$, etc.

Decimal #	Binary #
1	1
2	10
3	11
4	100
.	.
.	.
.	.
16	10000
17	10001
18	10010
.	.
.	.

Question 9 is on adding two binary numbers. In addition the following must be true: $0+0=0$, $1+0=1$, and $1+1=10$ (one, zero) and $1+1+1=11$ (one, one). As in decimal, addition of larger numbers can be accomplished by "lining up" the places and successively applying the addition rules from right to left. Thus:

$$\begin{array}{rl} 1111 & \rightarrow \text{ Carries} \\ 101101 & \\ +\ 110100 & \\ \hline 1100001 & \quad \text{Answer a in Q9} \end{array}$$

When writing about more than one system of numbering, it is often convenient to designate which system we are using. This is done normally by designating the base, or radix of the system. The base is the amount of symbols used in a particular system. The decimal system has a base of 10 because there are 10 symbols. The binary system has two symbols therefore the base or radix is 2 (two). The following shows how to designate the base when systems are mixed:

$$101_{(2)}=5_{(10)}$$

This is read "one zero one to the base two is equal to five to the base ten."

A numbering system with a base or radix of 8 is called an octal system. In counting in this system the digits used are 0, 1, 2, 3, 4, 5, 6, and 7. Hence counting in the base 8 would proceed as follows: 0, 1, 2, 3, 4, 5, 6, 7, 10, 11, 12, and so forth. The answer to Q10 is c. The place values in this system (proceeding from the right, least significant digit) are units, eights, sixty fours, five hundred twelves, etc. The base eight number 523 means there are five 64s, two eights, and three units. Or the decimal value is $(5\times 64)+(2\times 8)+(3\times 1)=339$.

$$523_{(8)}=339_{(10)}$$

One of the reasons octal numbering is popular in the computer world is because it is easy to convert binary numbers to octal numbers. If a binary number's digits are arranged in groups of threes, and each "triad" is converted to an octal number, the result is an octal equivalent of the binary number. An example should help illustrate this point.

Suppose the binary number 1101110110 is to be converted to an octal number. First group the numbers in "traids" thusly:

(00)1 101 110 110

Notice that the extra zeros added to the left really won't change the value, they are there only for grouping purposes. Now convert each group of three to an equivalent octal value (same as converting to decimal because three binary digits can only count as high as 7) as shown here:

(00)1	101	110	110
↓	↓	↓	↓
1	5	6	6

If this procedure is correct then:

$$1101110110_{(2)} = 1566_{(8)}$$

To check your work, convert each number to its decimal equivalent and compare the results. To make the problem consume less space, first write the binary number. Next jot the binary place value above each single digit in the binary number, as follows:

(512)	(256)	(128)	(64)	(32)	(16)	(8)	(4)	(2)	(1)
1	1	0	1	1	1	0	1	1	0

Multiply the place value by the digit in that place (either 1 or 0) and add the results:

$$(512 \times 1) + (256 \times 1) + (128 \times 0) + (64 \times 1) + (32 \times 1)$$
$$+ (16 \times 1) + (8 \times 0) + (4 \times 1) + (2 \times 1) + (1 \times 0) =$$
$$512 + 256 + 64 + 32 + 16 + 4 + 2 = 886$$

Therefore: $1101110110_{(2)} = 886_{(10)}$.

Now do the same for the octal number, as follows:

(512)	(64)	(8)	(1)
1	5	6	6

Again multiplying the place value by the digit in the places you obtain the decimal equivalents:

$$(512 \times 1) + (64 \times 5) + (8 \times 6) + (1 \times 6) =$$
$$512 + 320 + 48 + 6 \qquad = 886$$

Therefore: $1566_{(8)} = 886_{(10)}$.

From the above example you can see that the following is proven:

$$886_{(10)} = 1566_{(8)} = 1101110110_{(2)}$$

Again this notation is used because it is easy to convert to a number and it is more easily read and written. By arranging a group of LEDs (corresponding to the binary information of interest in a computer) into groups of threes, the octal number can be easily read with a little practice, and written down with much more ease and accuracy than the binary number.

Hexadecimal is of interest for the same reason. If the binary groups are in fours instead of threes, four bits can be read as a hexadecimal number quite easily. Let's try the same binary number but group it in fours. First, however, let's count in hexadecimal or base 16. In hexadecimal there are 16 symbols which range from 0 through F as shown in Table 2-2.

Table 2-2. Comparison of decimal, hexadecimal, and binary.

Base 10 Decimal	Base 16 Hexadecimal	Base 2 Binary
0	0	0
1	1	1
2	2	10
3	3	11
4	4	100
5	5	101
6	6	110
7	7	111
8	8	1000
9	9	1001
10	A	1010
11	B	1011
12	C	1100
13	D	1101
14	E	1110
15	F	1111
16	10	10000
17	11	10001

Notice that four binary digits take you through a count of F in hexadecimal or 15 in decimal.

Now getting back to our binary number that we converted to octal. It was $110110110_{(2)}$. Let's arrange this number in groups of four and convert each "quatro" into its equivalent hexadecimal number.

$$\begin{array}{ccc} (00)11 & 0111 & 0110 \\ \downarrow & \downarrow & \downarrow \\ 3 & 7 & 6 \end{array}$$

If this really works, then:

$$1101110110_{(2)} = 1566_{(8)} = 376_{(16)} = 886_{(10)}$$

Having already checked binary and octal, let's use the same check on the hexadecimal number, as shown below:

$$\begin{array}{ccc} (256) & (16) & (1) \\ 3 & 7 & 6 \end{array}$$

Multiply the place value by the digit as before prior to adding the results:

$$\begin{aligned} &(256 \times 3) + (16 \times 7) + (1 \times 6) = \\ &768 + 112 + 6 \qquad = 886 \end{aligned}$$

Therefore: $376_{(16)} = 886_{(10)}$.

Question 11 asks for the octal equivalent of the binary number 110101. Arranging it in groups of three, you should be able to easily convert to 65 (answer b).

By knowing the place value, the decimal equivalent of $2E_{(16)}$ in Q12 can be found.

$$\begin{aligned} &(16 \times 2) + (E \times 1) = \\ &32 + 14 \qquad = 46 \end{aligned}$$

Grouping each binary number into "quads" in Q13 shows the following:

Answer a was $1111\ 0100_{(2)} = F4_{(16)}$
Answer b was $1010\ 0010_{(2)} = A2_{(16)}$
Answer c was $0111\ 0111_{(2)} = 77_{(16)}$

Therefore answer d is correct as there was no correct number.

Boolean algebra

In Boolean algebra, mathematical notation is different than in regular algebra. In regular algebra AB; $A \bullet B$; and $(A)(B)$ all mean the quantity A times the quantity B. In Boolean algebra these sequences mean the *logical operation*, AND. Thus AB, $A \bullet B$, and $(A)(B)$ mean the quantity *A AND* the quantity B. The answer to Q14 is therefore d, all of the above.

Also in Boolean algebra, all quantities are limited to only two states. When using Boolean algebra, the number of quantities a function can possess is limited. Normally in computers these values are a logic "1" state or a logic "0" state. Therefore a function can only be a "1" or a "0" at any one time.

A bar or vinculum is used in Boolean algebra to indicate an inversion, or NOT, function. Thus $\overline{A}$ is read "not A." If a function is designated as B on the input to an inverter, the output is $\overline{B}$, as in Fig. 2-4.

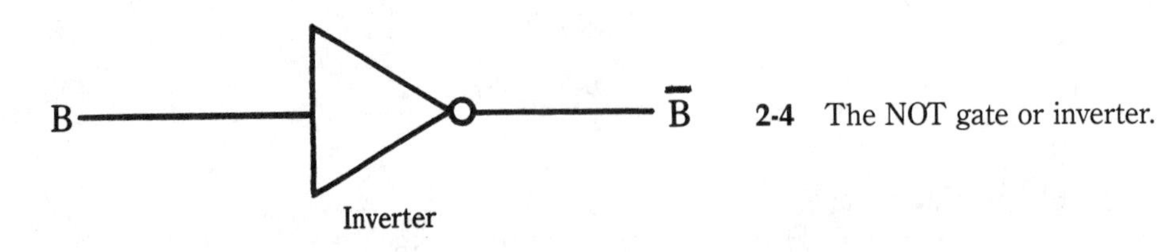

2-4 The NOT gate or inverter.

In Q15 the statement $A \bullet \overline{A}$ could be represented by a logic diagram as shown in Fig. 2-5.

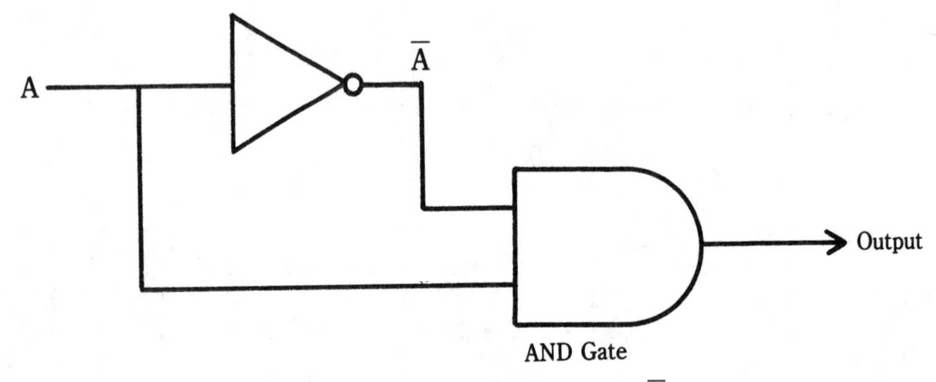

2-5 Logic diagram of the logic function $A \bullet \overline{A}$.

The output $\overline{A}$ will be ANDed with A. An AND gate has the truth table of Fig. 2-6. (A truth table is a table showing an output behavior for all possible combinations of input quantities.)

A	B	A·B
0	0	0
0	1	0
1	0	0
1	1	1

2-6 Truth table of the AND logic function.

In other words an AND function will be "1" only when all inputs are "1."

Now getting back to Q15, the inputs to the AND gate can never be "1" at the same time because if A is "1," $\overline{A}$ is "0," and if A is "0," A is "1." Therefore the output must be "0" at all times. We can therefore write $A \bullet \overline{A}$ = "0." The answer to Q15 is b, "0."

The regular algebraic symbol for addition was chosen in Boolean algebra to represent the logical OR function. Hence $A+B$ is read "the quantity A or the quantity B," or simply "A or B." A truth table for an OR function is shown in Fig. 2-7.

A	B	A+B
0	0	0
0	1	1
1	0	1
1	1	1

2-7 Truth table of the OR logic function.

Reduction of Boolean expressions is not covered on the Associate Certification Exam because the subject is not needed much in technical work. A few items for reduction are shown in Q16 for practice. All technicians should be aware of the methods used for reduction even if they are not required often.

With a few axioms of Boolean algebra and some practice, the equations in Q16 should be easy to reduce. Some Boolean algebra axioms are shown in Table 2-3.

Table 2-3. Axioms of Boolean algebra.

(1)	$1 \bullet 1$	=	1
(2)	$1 \bullet 0$	=	0
(3)	$1+0$	=	1
(4)	$1+1$	=	1
(5)	$A \bullet A$	=	A
(6)	$A+A$	=	A
(7)	$1 \bullet A$	=	A
(8)	$0 \bullet A$	=	0
(9)	$A+1$	=	1
(10)	$A+0$	=	A
(11)	$\overline{A}+A$	=	1
(12)	$\overline{A} \bullet A$	=	0
(13)	$\overline{\overline{A}}$	=	A
(14)	$AB+AC$	=	$A(B+C)$
(15)	$\overline{A} \bullet \overline{B}$	=	$\overline{A+B}$
(16)	$\overline{A}+\overline{B}$	=	$\overline{A \bullet B}$

Using these rules, reduce $AB+AC+A\overline{C}$, as follows:

From (14)→ $AB+AC+A\overline{C}=A(B+C+\overline{C})$
From (11)→ $=A(B+1)$
From (9)→ $=A(1)$
From (7)→ $=A$

Just for a check, let's construct a truth table for this expression to see if $AB+AC+A\overline{C}$ can actually be replaced by the single variable, A (Fig. 2-8).

A	B	C	$\overline{C}$	AB	AC	$A\overline{C}$	$AB+AC+A\overline{C}$
0	0	0	1	0	0	0	0
0	0	1	0	0	0	0	0
0	1	0	1	0	0	0	0
0	1	1	0	0	0	0	0
1	0	0	1	0	0	1	1
1	0	1	0	0	1	0	1
1	1	0	1	1	0	1	1
1	1	1	0	1	1	0	1

2-8 Truth table for the logic function $AB+AC+A\overline{C}$.

Notice that the column for A is identical to the last column. This means that $A = AB + AC + A\overline{C}$ and that our reduction was correct.

Now let's find the answers for Q16:

Q16 a. is $ABC + AB\overline{C} =$

From (14)→ $AB(C + \overline{C}) =$

From (11)→ $AB(1) =$

From (7)→ $AB =$

Q16 b. is $X(X + Y) =$

From (14) (backwards)→ $XX + XY =$

From (5) → $X + XY =$

From (14) and (7)→ $X(1 + Y) =$

From (9)→ $X(1) =$

From (7)→ X

Q16 c. is $(G + H)(G + K) =$

To simplify this one we need to show that the AND works like multiplication in algebra—that is that each term in the left parentheses is ANDed into each term in the right parentheses. To do this let's first let $A = G + H$, then:

$(G + H)(G + K) =$

$A(G + K) =$

From (14)→ $AG + AK$

Putting $G + H$ back in, for A, results in:

$(G + H)G + (G + H)K =$

From (14)→ $GG + GH + GK + HK =$

From (5)→ $G + GH + GK + HK =$

From (14)→ $G(1 + GH + GK) + HK =$

From (9)→ $G(1) + HK =$

From (7)→ $G + HK$

Q16 d. is $\overline{A}\overline{C} + \overline{A}C + BC =$

From (14)→ $\overline{A}(\overline{C} + C) + BC =$

From (11)→ $\overline{A}(1) + BC =$

From (9)→ $\overline{A} + BC$

Q16 e. is $\overline{A}B + \overline{A}\overline{B} + A =$

From (14)→ $\overline{A}(B + \overline{B}) + A =$

From (11)→ $\overline{A} + A = 1$

Q16 f. is $\overline{A}\overline{B}(A + B) =$

From (14)→ $\overline{A}\overline{B}A + \overline{A}\overline{B}B =$

rearranging

$(\overline{A}A)\overline{B} + \overline{A}(\overline{B}B) =$

From (12)→ $0 \bullet \overline{B} + \overline{A} \bullet 0 =$

From (2)→ $0 + 0 = 0$

Any time an OR input is "1" the output will be "1." Question 17 asks what an OR gate output will be with three inputs at "1," "0," and "0" respectively. The truth table shows that the output will be "1" hence answer c is correct.

In logic several combination gates are popular because of the physical structure; OR gates followed by inverting amplifiers were quite popular. When they were incorporated in one package, it became convenient to describe the package in terms of its total logical function—the OR followed by a NOT (inverter) function became a NOT-OR shortened to NOR. A NOR truth table is shown here in Fig. 2-9.

A	B	A+B	$\overline{A+B}$ = NOR
0	0	0	1
0	1	1	0
1	0	1	0
1	1	1	0

2-9 Truth table for the NOR logic function.

A NAND is a NOT AND functionally. The truth table appears in Fig. 2-10. This gives the answer to Q18.

A	B	A·B	A·B = NAND
0	0	0	1
0	1	0	1
1	0	0	1
1	1	1	0

2-10 Truth table for the NAND logic function.

Be careful when using the bar over the letters. Notice that $\overline{AB}$ does not equal $\overline{A}\overline{B}$. This can be shown by comparing the Fig. 2-11 circuits that give these two functions.

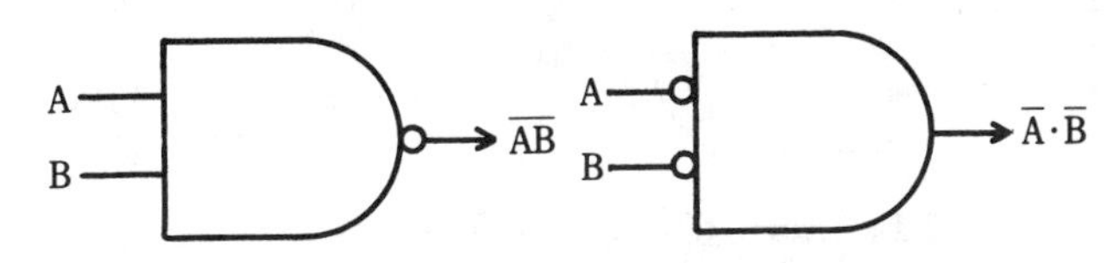

2-11 Logic circuit for $\overline{AB}$ and for $\overline{A} \bullet \overline{B}$.

For ease of comparison a truth table Fig. 2-12 can be used to keep track of what each output is for various combinations of inputs.

Column 3 simply shows the AND function, and *A* ANDed with *B*. Column 4 is found by inverting column 3. Column 4 is the output of the first circuit above. Columns 5 and 6 are simply inverses of *A* and *B* respectively. Column 7 is 5 and 6

1.	2.	3.	4.	5.	6.	7.
A	B	A·B	$\overline{A \cdot B}$	$\overline{A}$	$\overline{B}$	$\overline{A} \cdot \overline{B}$
0	0	0	1	1	1	1
0	1	0	1	1	0	0
1	0	0	1	0	1	0
1	1	1	0	0	0	0

2-12 Truth table showing that $\overline{AB}$ does not equal $\overline{A} \bullet \overline{B}$.

ANDed together, or in other words $\overline{A} \bullet \overline{B}$. Column 4 and column 7 are not the same for every combination of inputs therefore $\overline{A \bullet B} \neq \overline{A} \bullet \overline{B}$.

The answer to Q19 is a. An exclusive OR is similar to a regular OR except the output is excluded when both A and B are "1" at the same time. A truth table is shown here in Fig. 2-13.

A	B	A⊕B
0	0	0
0	1	1
1	0	1
1	1	0

2-13 Exclusive OR truth table.

Checking each answer in problem 15 against the truth table is one way to solve this problem. This can be done by substituting a circuit for each equation. Answer a is $A\overline{B} + \overline{A}B = F$. A circuit to represent this is shown in Fig. 2-14.

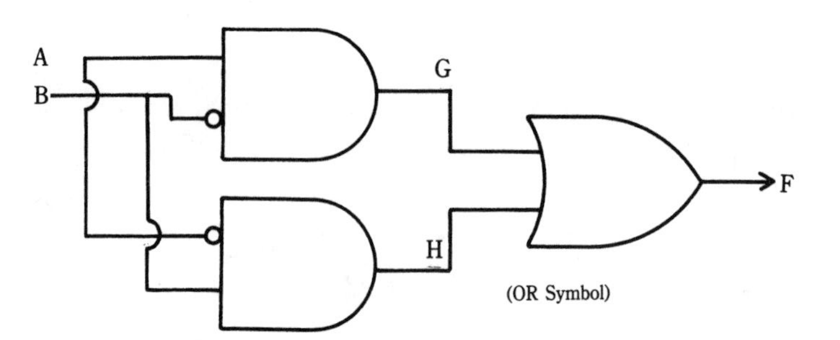

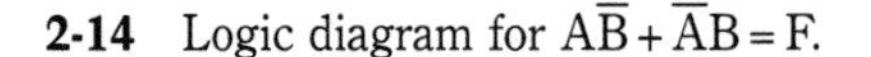
2-14 Logic diagram for $A\overline{B} + \overline{A}B = F$.

In the circuit of Fig. 2-14, when $A = 0$ and $B = 0$ then $G = 0$ and $H = 0$. G OR H then equals 0, which is F. If $A = 0$ and $B = 1$ then $G = 0$ and $H = 1$ and $G + H = 1$. If $A = 1$ and $B = 0$ then $G = 1$ and $H = 0$, therefore, $G + H = 1$ again. But if $A = 1$ and $B = 1$ then $G = 0$ and $H = 0$, making $G + H = 0$. Answer a for Q15, therefore, is correct. Answers b and c can be checked in a similar manner just to make sure they are wrong. A circuit for b is shown in Fig. 2-15.

The circuit for answer c would be as in Fig. 2-16.

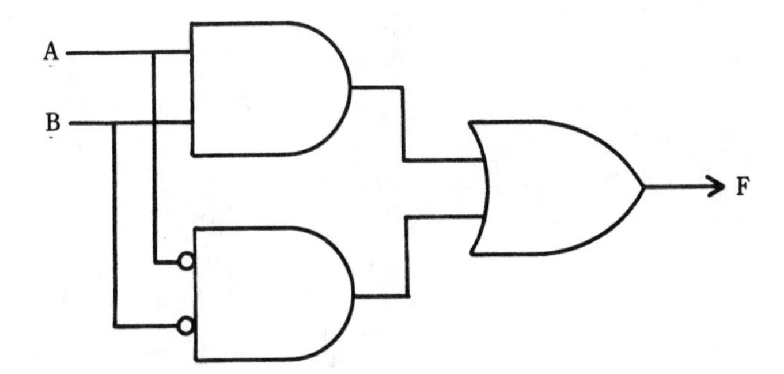

2-15 Logic diagram for incorrect answer b.

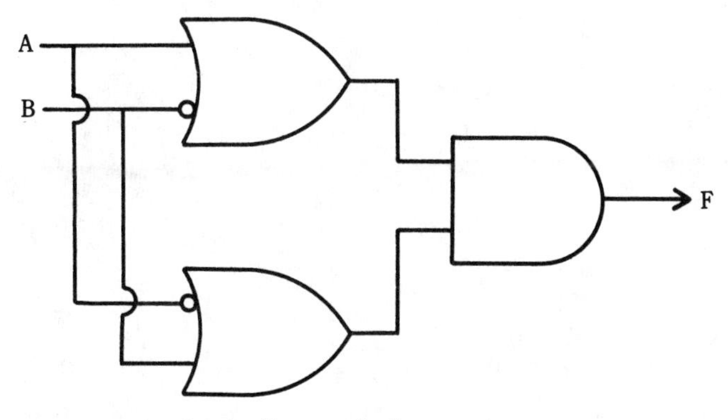

2-16 Logic diagram for incorrect answer c.

3 Electrical

THE MOST EXOTIC VIDEO GAMES, COMPUTERS, MILITARY FIRE CONTROL units, and satellite earth stations work with the same basic elements that have been used since the first radios initiated us into the science of electronics. Every technician must be familiar with these basic electronic components to understand how an electronic product works, and what, if anything, might be malfunctioning. Below is a quiz relating to these basic elements.

Quiz

Q1. How much current is flowing in the circuit of Fig. 3-1?

a. 1 amp
b. 0.1 amp
c. 10 amps
d. 100 amps

Q2. In the circuit of Fig. 3-1, how much power will be dissipated by the resistance?

a. 1 watt
b. 10 watts
c. 100 watts
d. 1000 watts

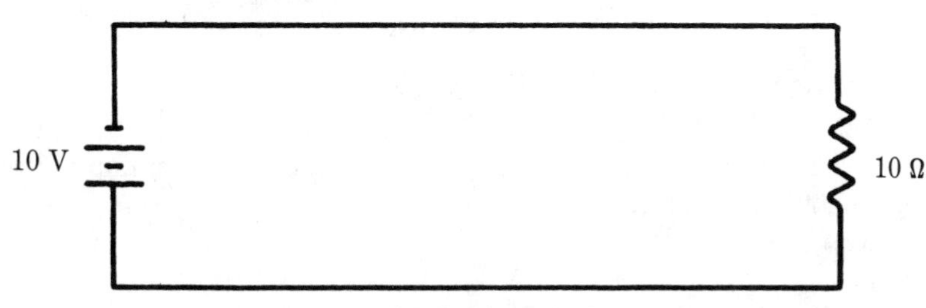

3-1 Circuit for problems in Q1 and Q2.

Q3. The pi-type filter in Fig. 3-2, C1, L1, and C2 will change the pulsating ripple voltage at *A* to ________ at *B*?

a. A 10-volt peak-to-peak waveform.
b. The same level waveform.
c. A 0.1-volt waveform.

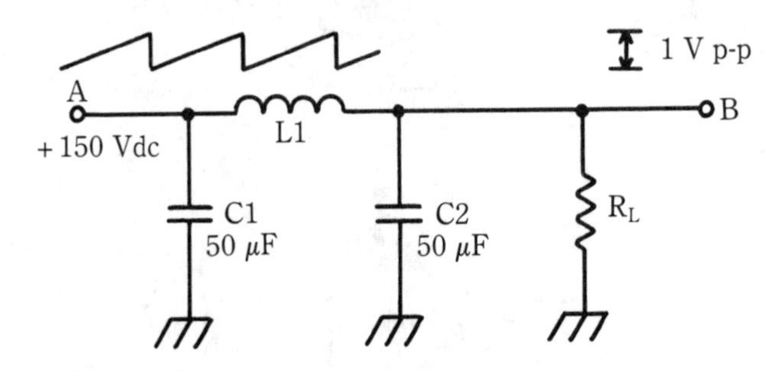

3-2 Pi-type filter for Q3, Q14, and Q15.

Q4. The voltage across R1 in Fig. 3-3 is:

a. 25 volts
b. 75 volts
c. 37.5 volts

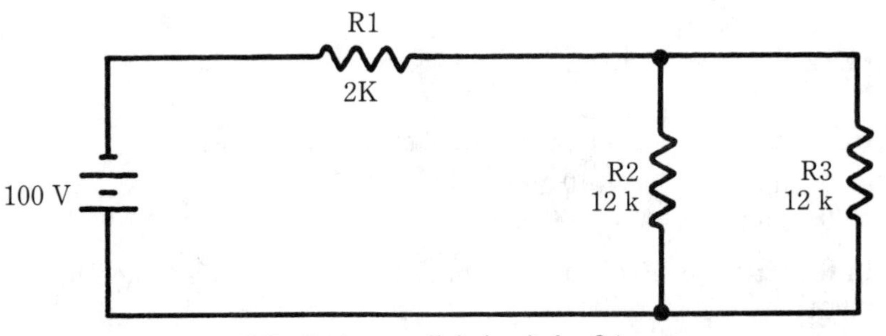

3-3 Series-parallel circuit for Q4.

Q5. The effective total capacitance across R_L in the circuit of Fig. 3-4 is:

a. 4 microfarads
b. 6 microfarads
c. 8 microfarads
d. 12 microfarads

Q6. If the circuit in Fig. 3-4 operates near 150 Vdc then what minimum voltage ratings should C1 and C2 have?

a. 75 volts
b. 150 volts
c. 300 volts

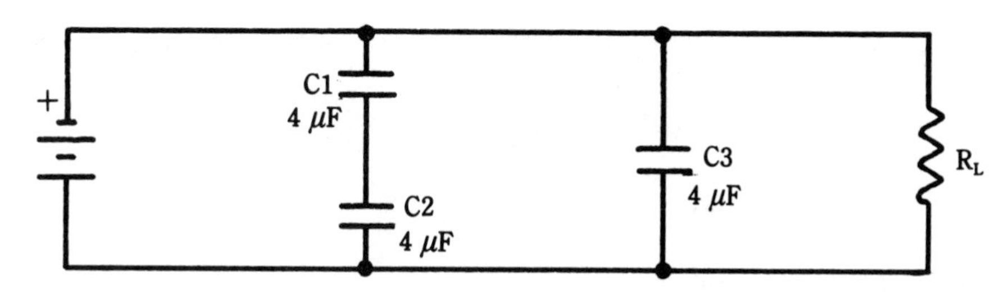

3-4 Series-parallel circuit for Q5 and Q6.

Q7. A resistor that changes value greatly with small changes in voltage across it is:

a. a thermistor.
b. a varistor.
c. a thyristor.

Q8. The values of two or more inductors in series or parallel combine like:

a. capacitors.
b. resistors.
c. neither.

Q9. A coil with the same length of wire but whose coils are spaced further apart will have:

a. less inductance.
b. more inductance.

Q10. Match the following:

a. inductor	x. current lags voltage
b. capacitor	y. current and voltage are in phase
c. resistor	z. current leads voltage

Q11. If the LC circuit in Fig. 3-5 is resonant at 500 kHz, how will it affect the incoming 500-kHz signal?

a. It will allow 500 kHz to pass through to the output but will resist or oppose lower or higher frequencies.

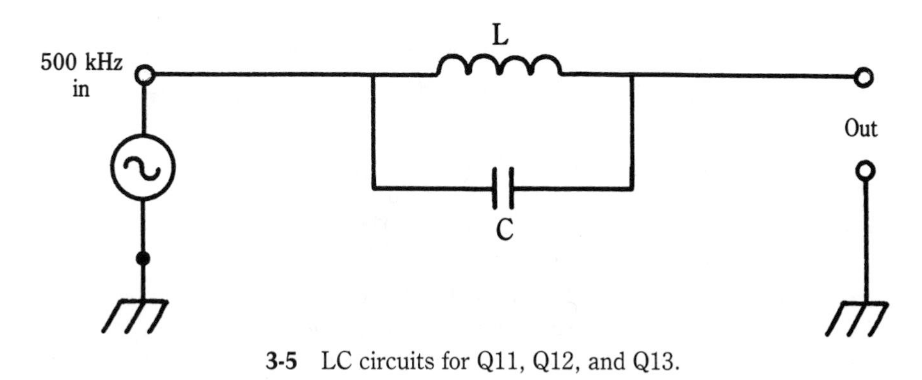

3-5 LC circuits for Q11, Q12, and Q13.

b. It will present a high impedance to 500 kHz but will allow higher or lower freqeuncies to pass to the output.

Q12. Calculate the capacitive reactance presented to 500 kHz in the circuit of Fig. 3-5.

a. if C = 0.001 microfarad	a. ________
b. if C = 0.000001 microfarad	b. ________
c. if C = 1000 microfarad	c. ________
d. if C = 0.00001 microfarad	d. ________

Q13. Figure the inductive reactance in Fig. 3-5:

a. if $L = 1$ henry
b. if $L = 50$ millihenrys
c. if $L = 50$ microhenrys

Q14. Looking back to Fig. 3-2, what is the capacitive reactance of C1 to 120 Hz?

a. 0.00265 ohm
b. 26.5 ohms

Q15. If, in Fig. 3-2, L1 is 2.6 henrys, what is its inductive reactance at 120 Hz?

a. 1.959 ohms
b. 1959 ohms

Q16. A resistor color-coded yellow, violet, silver is:

a. 0.47 ohm
b. 4.7 ohms
c. 470 ohms

Q17. What is the total reactance in the circuit of Fig. 3-6?

a. 16.3 ohms
b. 12 ohms
c. 260 ohms

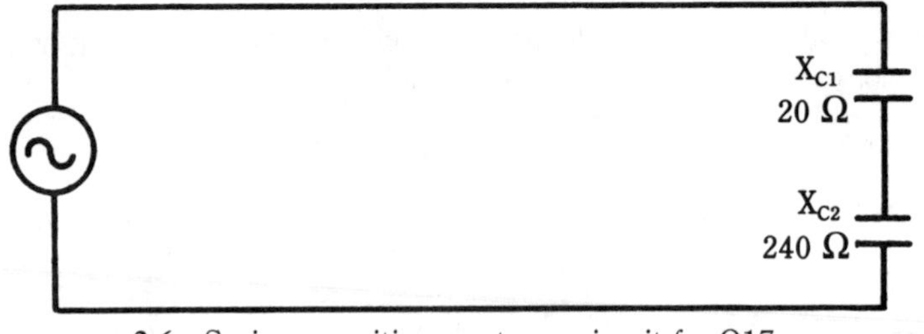

3-6 Series capacitive reactance circuit for Q17.

Q18. With an ac waveform of 100 volts on the primary and a turns ratio of 10 to 1, the secondary in Fig. 3-7 will have:

a. 10 volts and half the current of the primary.
b. 10 volts and 10 times the current of the primary.
c. 10,000 volts and 1/10000 of the current in the primary.

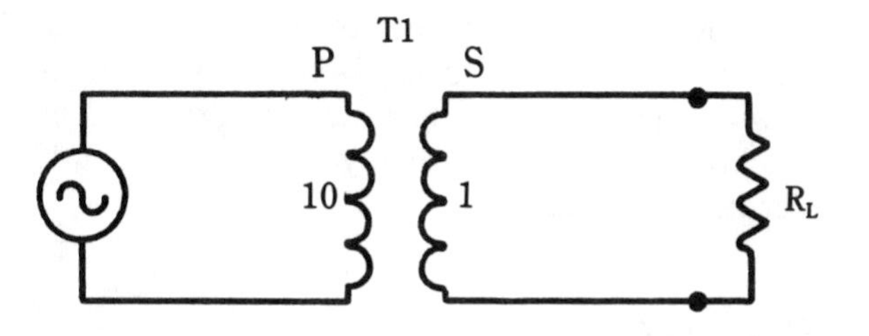

3-7 Transformer circuit for Q18.

Q19. If the waveform on the primary in Fig. 3-8 is 500 volts, what should you expect at A and B?

a. over 5000 volts at A, over 500 volts at B.
b. over 5000 volts at A, under 10 volts at B.

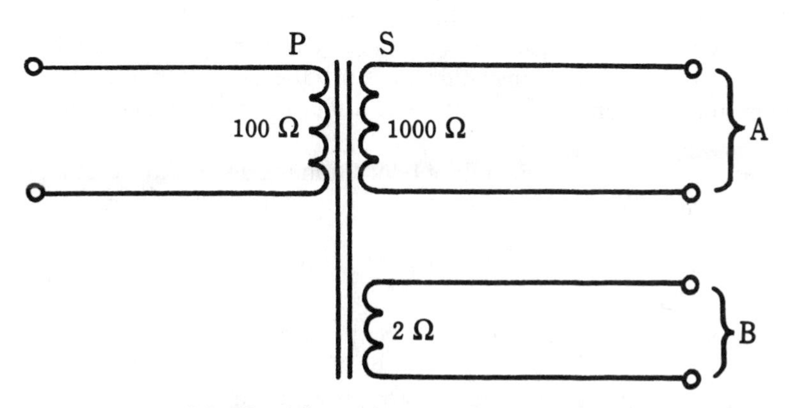

3-8 Partial transformer circuit for Q19.

Q20. What is the RC time of the circuit in Fig. 3-9?

a. 0.47 second
b. 0.0047 second
c. 47 microseconds

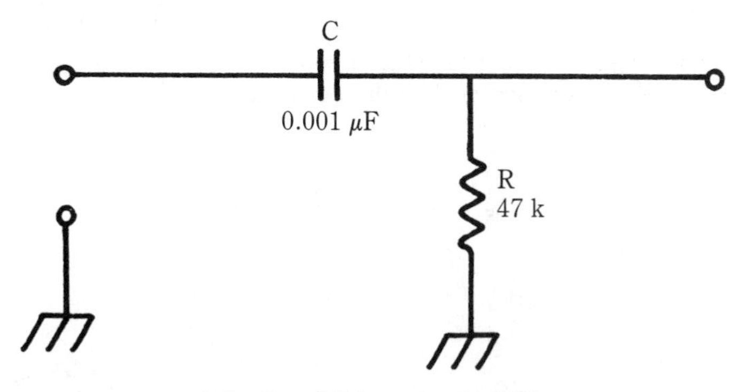

3-9 C and R in series for Q20.

Now that you have taken the quiz regarding basic electricity and the basic elements of electronic circuits, ask yourself if any of the questions were difficult. If you were not able to systematically work out the answers, or the answers weren't quickly evident to you, then you need to do further study before attempting the Certification Exam.

Electrical units

Several questions on the ETAI Certification Exams deal with basic electricity. They aren't all as simple as Q1, but they still test your understanding of current flow and the relationships between voltage, current, and resistance.

One ampere of electrical current is 6.25×10^{18} electrons flowing past a given point, per second. One volt is the practical unit used to identify the electromotive force needed to move 6.25×10^{18} electrons between two points. Instead of referring to that large number, we refer to that quantity as a *coulomb*. To put into perspective the amount of work done by a one volt charge moving a coulomb of electrons past a given point in one second, it is about 3/4 of a foot-pound of work, or one joule of work.

One *ohm* of resistance is the unit of opposition to current flow that allows only 1 amp of current to flow when moved by a 1-volt charge. In the formula

$$I = \frac{E}{R}$$

the amount of current in an electrical circuit equals the voltage divided by the resistance.

Some of the relationships you are introduced to in school never (or at least rarely) will be referred to after graduation. That's not true about Ohm's law, especially the power calculations. You will need to calculate the wattage of a resistor you intend to place in a circuit, or you will want to estimate the amount of excess current that might be the fault symptom in a circuit you are troubleshooting. The power formulas are:

$$P = E \times I; \; P = \frac{E^2}{R}; \; P = I^2 R$$

Or, you can transpose in problems where you have the amount of current, resistance, or voltage when you know the amount of power:

$$E = \frac{P}{I}; \; I = \sqrt{\frac{P}{R}}; \; E = \sqrt{PR}$$

$$\text{or: } I = \frac{P}{E}; \; R = \frac{P}{I^2}; \; R = \frac{E^2}{P}$$

Some technicians have difficulty remembering Ohm's law unless their job calls for its constant use. One helpful method is to name the elements: Eagle, Rabbit, and Indian. In the basic Ohm's law formulas and power formulas (excepting the easiest one: $P = EI$) the Eagle is always in the air while the Rabbit and Indian are on the ground:

$$I = \frac{E}{R};\ R = \frac{E}{I};\ E = IR;\ P = \frac{E^2}{R};\ P = I^2R$$

In calculating I, E, and R, when given the other two factors, most technicians have little difficulty with simple problems as in Q1 and Q2. When the circuit becomes complex and divides into multiple paths the solutions become more difficult. However, by dividing the circuit into subcircuits and working out the paths one portion at a time, and by knowing how to work Ohm's law formulas, you should have no difficulty with these questions. The answer to Q1 is a, and Q2 is b.

Wattage problems seem to confuse some technicians. The question of how to add power dissipations in series and parallel gets mixed up with resistance or capacitance in series or parallel. Remember that wattages add, whether in parallel or in series.

Capacitors

In Q3 you are introduced to the most common power supply filter arrangement used in electronics. If we use an oscilloscope, we should see some small ripple voltage floating atop the +150 Vdc (which was rectified from the 117-Vac line). The filter combination reduces the ripple. *C*1 attempts to smooth out the rectified pulsating dc that is produced by the rectifier by acting as a giant reservoir for electrons. Any increase in voltage at *A* must first push a lot of electrons into *C*1 before the voltage can build to its full potential. Likewise, when the voltage attempts to decrease at A, any change will immediately be opposed by *C*1 because it contains a vast reservoir of electrons. *C*1 can't do it all, though, so *C*2 is added as insurance and acts in the same manner as *C*1.

In between *C*1 and *C*2 is *L*1, a coil that also acts as a "shock absorber." The coil works on the ripple in an opposite manner as the filter capacitors—where current *leads* the voltage in the capacitors (current flows in until the voltage is at its high point), the current *lags* the voltage in the inductor coil. Before any increase in voltage is impressed through *L*1, the current must trickle through the windings. As the current attempts to increase or decrease through the coil windings, it produces a back-emf that opposes that change, thus slowing the current. The current then lags the voltage across the coil. With the capacitors opposing change in one manner (current leading voltage) and the coil opposing change in an opposite manner (current lagging voltage), yet both opposing any changes, the ripple is smoothed to a very low level (dc). c is the best answer.

Combination circuits

Question 4 can be solved by calculating the combined resistances of *R*2 and *R*3 prior to trying to determine what portion of the 100-volt supply is dropped across *R*1. The formula for parallel resistance is:

$$R_T = \frac{R1 \times R2}{R1 \times R2}$$

However, where equal resistors are in parallel it is easier and faster to divide one of the equal resistors in half, in this case 6 kilohms. Then you can figure out the current flow using Ohm's law:

$$I = \frac{100}{6\text{ k} + 2\text{ k}};\ \mathrm{I} = \frac{100}{8\text{ k}};\ \mathrm{I} = 12.5\text{ mA};\ E = IR;$$

$$E = 0.125 \times 2000;\ \mathrm{E} = 25\mathrm{V}$$

It is easier to see that *R*2 and *R*3, totaling 6 kilohms, are equal to 75 percent of the total resistance. Because that is true, then 75 percent of the voltage will be dropped across *R*2 and *R*3, leaving 25 percent, or 25 volts to be dropped across *R*1.

Series and parallel capacitors are combined in Q5. In parallel capacitors the available plate area is increased and thus the electrostatic field area is increased. The opposite occurs in series. The electrostatic field must divide itself and attempt to attract or repel electrons on two "fronts." It is the same as increasing the thickness of the dielectric between the plates. As such the total capacitance will be less than the smallest capacitor, or if the two capacitors are equal, it will be 1/2 that of one of them. Don't get this confused with capacitive reactance. X_C and C are two different things.

The working voltage of combined capacitors is of concern in Q6. Because connecting two capacitors in series is in effect increasing the thickness of the dielectric, and the dielectric determines the working voltage the capacitor is able to withstand, then putting two capacitors in series allows you to add the maximum voltage ratings, two 75-volt rated capacitors serving in place of a single 150-volt capacitor.

Inductors and other components

Many terms used in electronics simply must be remembered. The name for various types of parts must be fixed in your mind. In Q7 you should eliminate answer a because that resistor, a thermistor, by its name, is affected by temperature. A *thyristor* (answer c) is a semiconductor, like an SCR, that functions similarly to a thyraton tube, turning on after a certain level of trigger voltage is reached, then not turning off until the power source is removed. A varistor, b, is the correct answer to Q7.

L = L1 + L2 + L3. Inductors in series add like resistors in series. Two coils in series act as one, with the combined inductance of both. Inductors in parallel combine the same as resistors in parallel. Two equal inductances in parallel equal only

1/2 the value of each one of them. Inductive reactances also combine like resistors. The answer to Q8 is therefore b.

It is the interaction of the magnetic fields surrounding the coil wire that produces the inductance. If you move magnetic fields away from each other, the attraction or repulsion is less between adajacent windings, thus putting wider spacing between the wire of a coil reduces the inductance; answer a in Q9.

In an inductor, a voltage pulse is impressed across the coils. Due to the back-emf caused by the interaction of the magnetic fields surrounding each turn of the wire, the current is opposed or slowed. Thus the current lags the voltage; a therefore matches x in Q10.

For b, the current fills the capacitor's reservoir—its wide plates—before the capacitor's electrostatic field reaches its maximum level; thus, the current leads the voltage in a capacitor, so b matches z.

For c, current and voltage are in phase in a resistor—there is no lead or lag, so c matches y.

Figure 3-10 gives comparative voltage, current, and power waveforms for inductors, capacitors, and resistors.

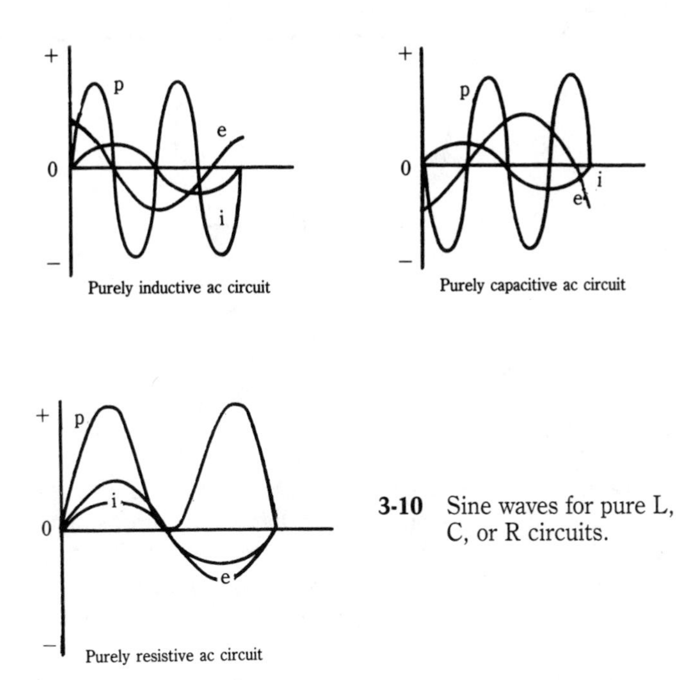

3-10 Sine waves for pure L, C, or R circuits.

Many incorrect answers on Certification Exams have been associated with circuits, such as the LC circuit in Q11. If the circuit is resonant at 500 kHz, then the inductive and capacitive reactances are equal at that frequency. Above 500 kHz, the capacitor will have less reactance and will tend to allow higher frequencies to pass through. The X_C will be greater at frequencies lower than 500 kHz. As f gets smaller, X_C gets larger:

$$X_C = \frac{1}{2\pi fC}$$

The capacitor tends to retard or limit frequencies below 500 kHz. The coil acts in an opposite manner allowing anything lower than 500 kHz to pass easily, but opposing any frequencies above 500 kHz. As a result the circuit acts to impede frequencies in the 500-kHz range only. It could be classified as a 500-kHz trap.

To better understand how the combination works, let's hook it up parallel to the signal or across the line, rather than in series with the signal, as in Fig. 3-11.

The circuit still presents an impedance to frequencies near 500 kHz, but instead of impeding them it presents a barrier that keeps the 500-kHz signals from being passed to ground. Highs are passed through the capacitor to ground, and lower frequencies pass through the coil to ground. Those near 500 kHz can be retained and passed right on through $R2$ and to the output.

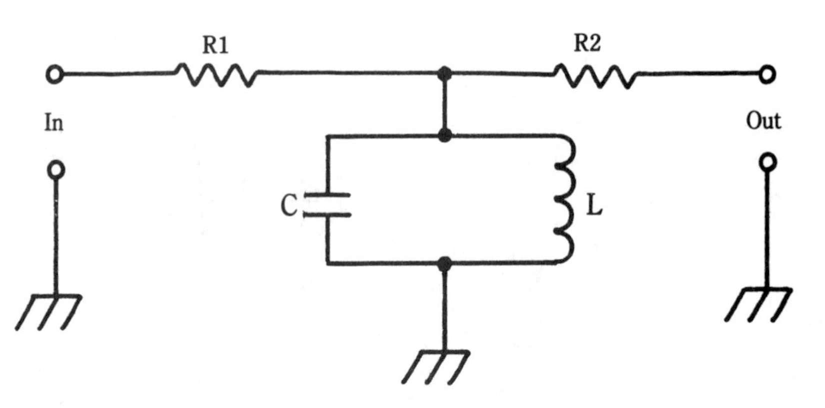

3-11 LC tuned circuit.

Reactances

If the working of the formula for X_C in Q12 is difficult for you, the best way to solve the problem is to set up a dozen or more hypothetical equations using some different values and work them until you are comfortable with it. It is unnatural to deal with such small and large numbers, but with a little practice you can master them. After you work one, you only change one number in your calculator to get the answers to the others.

The same discipline of working such equations until you have mastered them is needed to solve the inductive reactance problems of Q13. The relationship of a large coil and a small coil to the lower end of the radio-band frequencies (500 kHz) is important also.

You need to work out the answer for Q14 to make the right choice. Some people attempt to pass the Certification Exams by guessing, but it is virtually impossible. You must know how to solve the problems.

Question 15 more firmly implants the effect of a coil on the circuit at different frequencies. It shows why a large choke coil is necessary at ac line frequencies.

Miscellaneous problems

Don't forget the color codes that indicate tolerances, and the below-10-ohm band indicators. In the third, or multiplier band, if the color is silver, multiply by 0.01, and if it is gold, multiply by 0.1 for the correct value.

Reactance of capacitors adds like resistance—therefore 240 ohms and 20 ohms = 260 ohms in Q17. The question doesn't ask about the combined capacitances, which would be less than the value of the smaller capacitor.

Transformers inductively couple power from the primary to the secondary. The voltage coupled is directly proportional to the turns ratios. The current transferred is inversely proportional. The correct answer for Q18 is b.

The resistance of the wire in separate windings of a transformer is not directly related to the turns or the voltage. However, if the wire sizes are relatively close and the resistances are not close, as in Q19, the expected voltages can be approximated.

RC time is used in Q20 and on all Certification Exams. One RC time is simply $R \times C$, R is in ohms and C in farads. The decimal point and large number of zeros makes the math tough, but you must be able to work this type of problem. Also you need to know that in 1 RC time, the capacitor charges to 63 percent of the applied voltage and that it takes 5 RC times to charge to 99 percent, and 10 RC times to be considered fully charged. Discharging takes the same time, 1 RC time to discharge to 37 percent of the applied voltage. (Notice we did not say "1 RC time to discharge *to* 63 percent.")

Additional reading

Basic Electricity, Bureau of Naval Personnel, Dover Publishers, 1962.

Faber, Rodney B.: *Applied Electricity and Electronics for Technology*, John Wiley and Sons, 1978.

Horn, Delton T.: *Basic Electronics Theory—with projects & experiments*, TAB Books, 1981.

Kaufman, Milton and J.A. Wilson: *Basic Electricity: Theory and Practice*, McGraw-Hill, 1973.

Mandl, Matthew: *Fundamentals of Electric and Electronic Circuits*, Prentice-Hall, 1964.

4
Electronic components nomenclature

THIS CHAPTER COVERS THE TERMINOLOGY USED TO DESCRIBE ELECTRONIC component parameters.

Quiz

Q1. The resistor in Fig. 4-1 is:

a. A 47-kilohm resistor, 5 percent tolerance.
b. A 4700-ohm resistor.
c. A 470-ohm resistor.
d. A resistor that has an actual measurement within 1 percent of the marked value.

Q2. The wattage of the resistor is:

a. $^1/_4$ watt
b. 10 watts
c. 5 watts
d. Cannot be ascertained from the markings.

Q3. Color coding of a 5 percent, 6.8-ohm resistor will be:

a. green, purple, black, gold
b. blue, grey, silver, gold
c. blue, blue, silver, gold
d. blue, grey, gold, gold

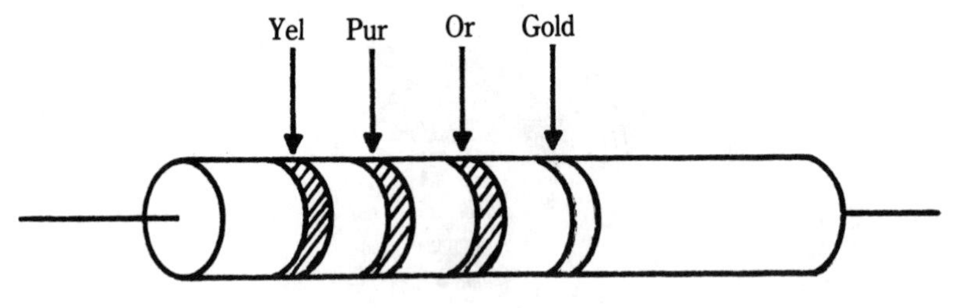

4-1 Resistor (see Q1 and Q2).

Q4. The resistor in Fig. 4-2 is:

a. marked wrong—should be marked R4.
b. a safety component and should be replaced with the exact value part.
c. one that might vary in size.
d. part of a multiresistor pak.

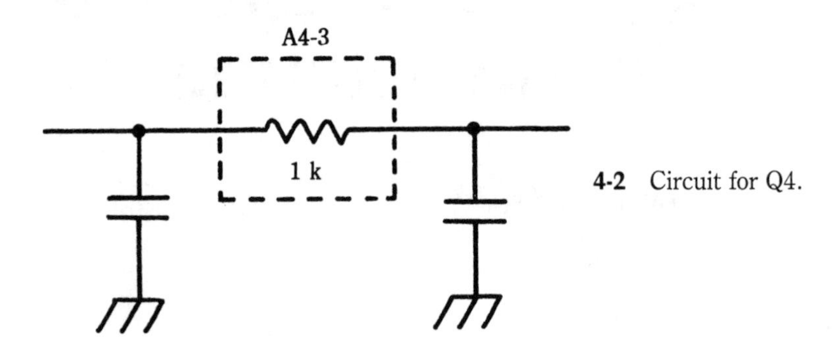

4-2 Circuit for Q4.

Q5. A disk capacitor with the marking 47 k will most likely be:

a. a 47-picofarad capacitor.
b. a 0.047-microfarad capacitor.
c. a 47-kilohm resistor that looks like a capacitor.
d. a 47-microfarad capacitor.

Q6. In Fig. 4-3, which symbol shows the standard transistor lead configuration?

a. 1 c. 3, 4, and 5
b. 2 d. There is no standard.

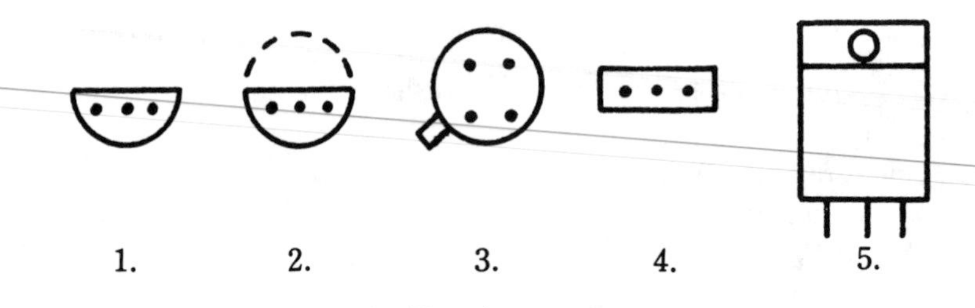

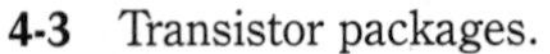
4-3 Transistor packages.

Q7. Figure 4-4 represents:

a. a bipolar transistor.
b. an SCR.
c. an FET.
d. a Darlington transistor.

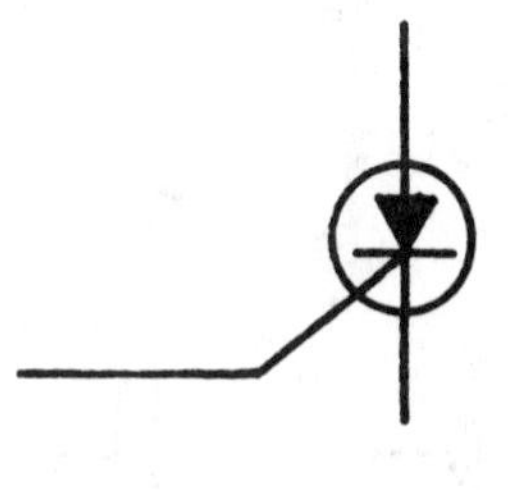

4-4 Solid-state component for Q7.

Q8. If V_B is 20 volts, $R1$ is 500 ohms, and $R2$ is 500 ohms, what will be the voltage across $D1$ (Fig. 4-5)?

a. 10 volts
b. 11.2 volts
c. 20 volts
d. impossible to determine from the information given.

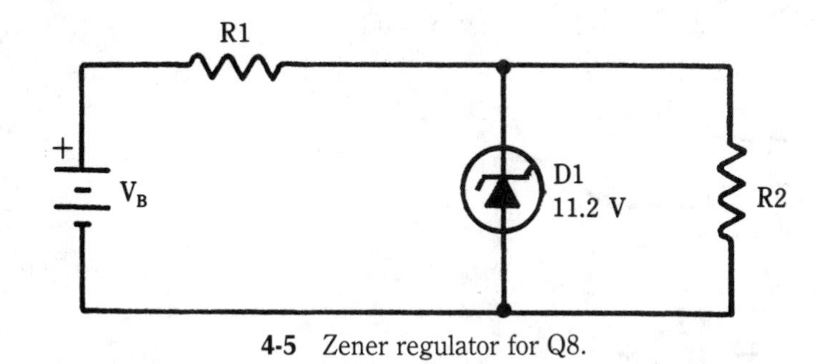

4-5 Zener regulator for Q8.

Q9. In Fig. 4-6, which leg is the drain?

a. a
b. b
c. c

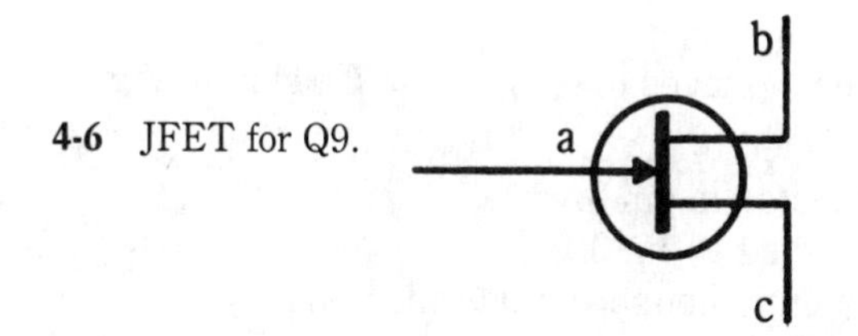

4-6 JFET for Q9.

Q10. The component in Fig. 4-7 is a wire with a black pencil-lead-appearing cylinder not directly connected, but loosely slipped over the wire. Which symbol represents it?

a. a   c. c
b. b   d. d

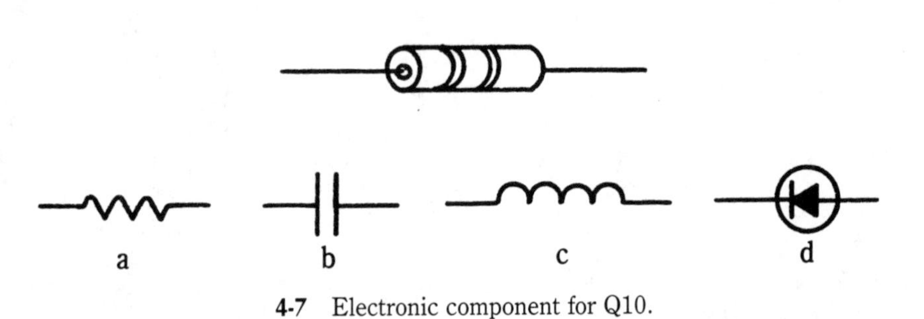

4-7 Electronic component for Q10.

Q11. Identify the symbols in Fig. 4-8.

a. AND gate, NOT gate, Exclusive-OR gate.
b. AND gate, NOR gate, Exclusive-NOR gate.
c. NAND gate, diode rectifier, Exclusive-OR gate.
d. AND gate, NAND gate, NOR gate.

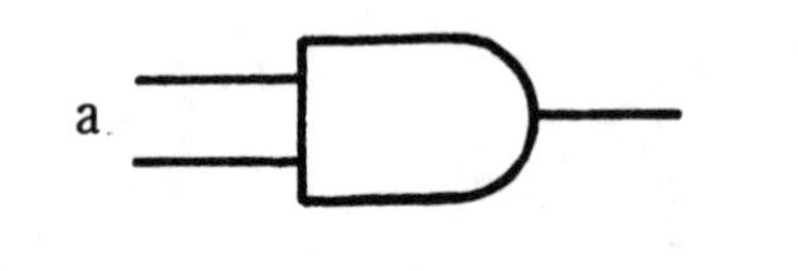

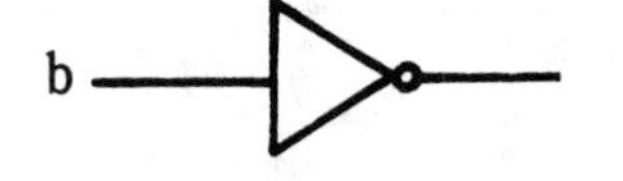

4-8 Logic symbol to be identified in Q11.

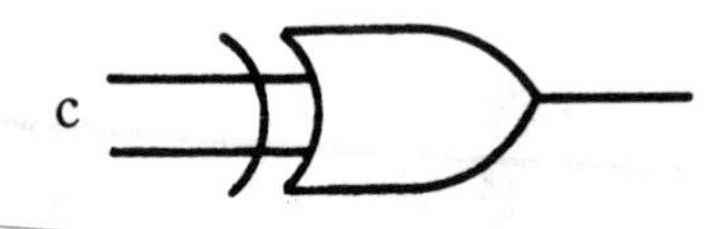

Q12. Power supply voltages to operate the *U*4 chip in Fig. 4-9:

a. are connected to pin 1.
b. are connected to pin 2.
c. are connected to pin 3.
d. are frequently not shown on schematics.

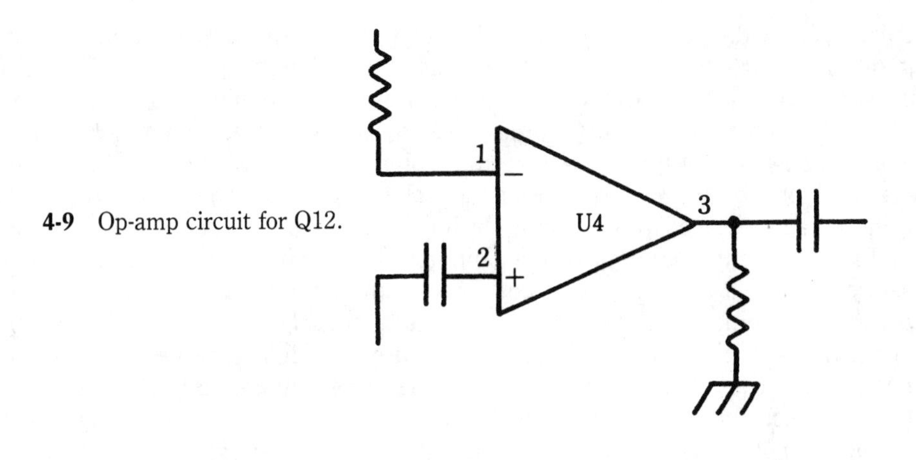

4-9 Op-amp circuit for Q12.

Q13. The component in Fig. 4-10 is:

a. a beam-power pentode.
b. a tetrode section of a multisection tube.
c. a gas-filled dual tetrode rectifier.
d. a triple diode/tetrode tube.

Q14. Match the following:

(1) RFC	(m) center conductor
(2) coax	(n) relay coil
(3) HOT	(o) flip-flop
(4) latch	(p) radio-frequency coil
(5) RL	(q) horizontal output transformer
(6) CRT	(r) kinescope
(7) TO-5	(s) round chip-package type

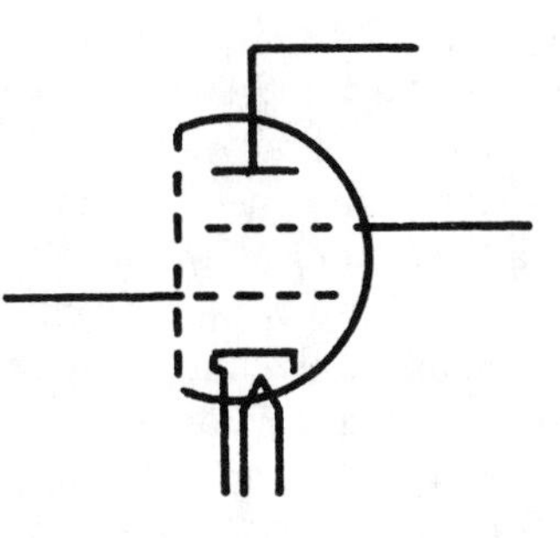

4-10 Electronic component for Q13.

Electronics industry standards

Any study of standard notation and symbology in the electronics industry is a futile experience. About the time transistor circuitry took over from tube technology, any serious efforts by manufacturers or their standards associations to maintain standards appear to have disappeared. Tube types, Morse code, and resistor color-code bands are just about the only items you can be sure of. Because of this giant problem, technicians spend large amounts of time, needlessly verifying the size and function of electronic components. Unmarked components are not a serious problem in manufacturing. For instance, a design modification might call for a change of a diode for frequency, power, or voltage reasons. The production technicians might then need to know that the "blue" colored diode should always be removed and replaced with the new "yellow" diodes. To the servicer in the field, undisclosed identification of components and arbitrary notation and symbols on service data is a nightmare that costs over a million dollars worth of time each *month*! The format of schematics and item number assignment for various components varies from manufacturer to manufacturer. It might even be different from one product to another that is produced by the same manufacturer. Because of this problem, technicians can only do their best and attempt to recognize components by the symbols that are "most prevalent" during the present period of time.

Resistors

Resistor color codes have traditionally been best remembered by technicians by using a phrase to remember the color sequence from 0 to 9. A helpful mnemonic is: "Better Be Right Or Your Great Big Venture Goes Wrong." Better stands for Black, which is the color used for zero; Brown represents 1; Red, 2; Orange, 3; Yellow, 4; Green, 5; Blue, 6; Violet, 7; Grey, 8; and White, 9. In Q1 the resistor shown in Fig. 4-1 has the 1st significant digit Yellow or 4. The second is Purple, or 7; and the multiplier—or better said—the number of zeros following the 2 significant digits is Orange, or 3. It is 47,000 ohms, or 47 kilohms. Notice that the colors go from Black to White in a natural progression. That helps you if you are a little rusty and seldom have need to identify resistor sizes. Also be aware that the color code works only above 9 ohms. Other rules are needed for smaller size resistors (under 10 ohms).

In Q2, d is correct. Carbon resistors still are made in physical sizes that are standard: 1/2-watt, 1-watt, and 2-watt sizes. In micro electronics work a 1/4-watt resistor is generally identifiable by physical size. Wire-wound 2-, 5-, and 10-watt resistors have been identifiable in most cases by physical size. The over-2-watt resistors often have the wattage printed on them. Wattage of non-flammable resistors are not marked. They are generally smaller in physical size than carbon or wire-wound resistors for the the same wattage values.

Under 10-ohm resistors present some difficulty for technicians. They are frequently used in power amplifier and power supply circuitry where heat causes discoloration of the color bands. For instance, a Blue, Grey, Silver, Silver might appear to have four bands of the same color after some use. A resistor value of 1 to 10 ohms uses a Gold third band. Gold means divide by 10. (When it is the third band rather than the fourth band, it is a divider, not the tolerance indicator.) So, if the first band is Red and the second is Red and the third is Gold, you count the Red closest to an end of the resistor as a 2. The second significant figure indicator is Red, therefore also a 2. That is 22 (twenty-two). Divide by 10 and the value is 2.2 ohms. The fourth band means the tolerance, or how far the resistor might actually measure from the indicated value. No fourth band means it might vary up or down as much as 20 percent; Silver means 10 percent and Gold 5 percent. In the early days of radio 20-percent resistors were common. In TV 10 percent was most often used and in digital circuits anything with tolerances more than 5 percent is rare, with 1-percent resistors encountered frequently. A Blue, Grey, Gold, Gold resistor as in Q3, is a 6.8-ohm resistor and might actually measure anywhere from 6.46 to 7.14 ohms.

The stitched box surrounding the 1-kilohm resistor in Q4 indicates that it is part of a multiresistor pak. The pak can be identified as A4 and the Fig. 4-2 section will be the third resistor from the dotted end. The resistor paks invariably contain resistances of the same value and are molded into shapes identical to IC chips—including 14-pin dip (dual-in-line packages). IC dip packages are usually black, resistor paks are usually blue or white.

Capacitors

Question 5 asks about a 47-kilohms marked capacitor. The answer b is correct. The small physical size of this type of capacitor forced manufacturers to find a way to identify the part with less digits. So the micro-microfarad base quantity was agreed on and the k multiplier then referred to a thousand $\mu\mu$F. Then 47 kilohms is 47,000 micro-microfarads, or 0.047 microfarad. The half-dozen or so early attempts at standard capacitor identification such as: Body-End-Dot; Color Bands; Arrow Boxes; Edge Dots, and so forth are not used anymore. Most often encountered capacitors have the sizes printed on them as well as the breakdown voltage. However, lately, more capacitors are seen in electronics products with secret identification. This requires using the circuit location and service data to determine the correct value.

Semiconductor devices

Question 6 is a little misleading because it asks for a standard that does not exist. Since the earliest days of transistor circuitry, there has been no standard lead configuration adopted by the manufacturers. Those parts that do contain the E, B, C locations marked on them, are a blessing for working technicians. If these aren't on the part, which is the most common condition, then manufacturers reference numbers must be consulted, or the replacement component package might have an outline drawing showing the lead locations. Inserting a transistor in a circuit the wrong way will destroy the transistor, and perhaps others along with it. Soldering it in more than once will tend to ruin the PC board traces. The best procedure is to make sure each lead is in the proper place, even though this takes valuable time and is tedious.

Question 7 asks you to identify the SCR symbol. You also should be prepared to identify gate, anode, and cathode and know the SCR function. The SCR symbol is one that has not varied from its original appearance. It can however be confused with the triac or the diac (Fig. 4-11).

In Q8, answer a is correct. The voltage source (20 V) will divide equally

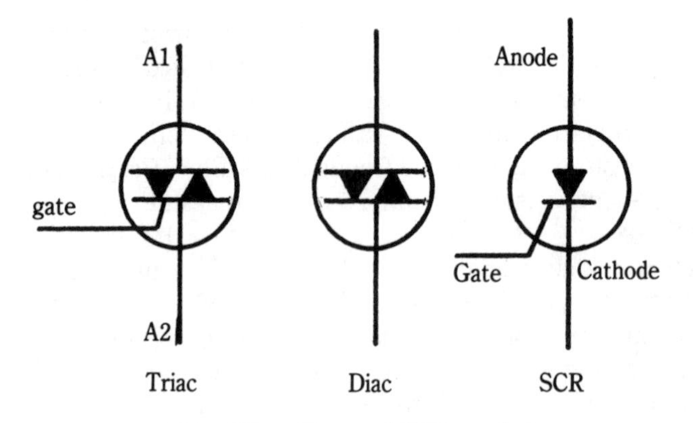

4-11 Triac, diac, and SCR symbols.

between the resistors leaving +10 volts from the D1 cathode to the anode. D1 will serve no purpose unless some circuit function should cause the voltage to rise to over 11.2 volts at the D1 cathode. Then D1 will conduct, keeping the voltage at 11.2. Notice the zener symbol; it is the symbol must commonly used. A diode symbol with a small "z" inside the circle is also used. You also will see zeners without the circle enclosure, and you will find them identified as CR; D; Z; X, and other part numbers.

Field-effect transistors (FETs) of all types are included in the Certification Exams. Know their advantages, their symbols, and how they are used. The answer to Q9 is that the JFET or junction-field-effect transistor is ordinarily drawn with the drain at b. Fortunately most schematics do show Source(s); Gate (g) and Drain (d) locations. JFETs are covered in chapter 9.

Many semiconductor circuits might need a small amount of inductance inserted, after the printed circuit board is designed, to eliminate parasitics. The drawing in Q10 (Fig. 4-7) is a ferrite sleeve that serves as an rf choke in rf tuners, horizontal output stages in TV sets and many other circuits. It is used to suppress unwanted rf oscillations, harmonics, etc. There is no identification and in a circuit it might have the appearance of a diode with the cathode band obliterated, or it might seem to be a burned resistor.

Logic symbols

More details of logic appear later in this book; you should familiarize yourself with these components. In Q11, a is correct. All CETs should be familiar with the military symbols for gates. These are now the most commonly used symbols in digital work. IEEE Std 91-1984 might replace these symbols. Examples are shown in Fig. 4-16.

Operational amplifiers

In the operational amplifier shown in Fig. 4-9 of Q12 answer d is correct. Where the product uses multiple chips that derive power from a common power supply, the supply pins on the chips might not be shown, merely the signal pins are shown. Sometimes the power supply pins are shown in the schematic legend as a note.

Other electronic components

Multisection tubes like the one in Q13 might be located far from each other, schematically. To show that a section is a part of a dual or triple envelope, the stitched line is used on one side of the circle. Four elements in a tube (not counting the filaments) make it a tetrode.

The items listed in Q14 as 1 through 7 are commonly used acronyms or words and you should relate them to the m through s names with little trouble. There are dozens of other commonly used acronyms in electronics. The more familiar you are with the common symbols and usage, the easier your technical work will be. Much of the terminology and present-day usage can only be gained by experience.

International standard notation

Many years ago it became evident that the method of identifying electronics components and noting component values had some problems. The problems were primarily encountered by service technicians but also, to a lesser degree, by engineers and others who work with electronic circuitry.

Figure 4-12 shows the most common method of schematically identifying the size of the capacitors. A severe problem is created when the decimal point in A or C is an original typographical omission, or if it gets lost in later reproductions or photocopies of the print. The decimal point takes up as much space as any other character and on small capacitor packages might be rubbed off or easily undetected. It is also possible that 0.1 μF is written 10 nF, or more commonly: 100,000 pF, or 100 k. Take your pick!

A more positive method of identification is the IS, or International Standards as adopted by ISA. Here in Fig. 4-13 are the same three capacitors of Fig. 4-12 with IS identification.

In Fig. 4-13, item A still has the space problem, but there is no decimal point. In item B, the F is dropped for farad because capacitors always have farads as a unit. In B one character is eliminated to conserve space without sacrificing understanding. In C four characters are removed and the possibility of a decimal error is

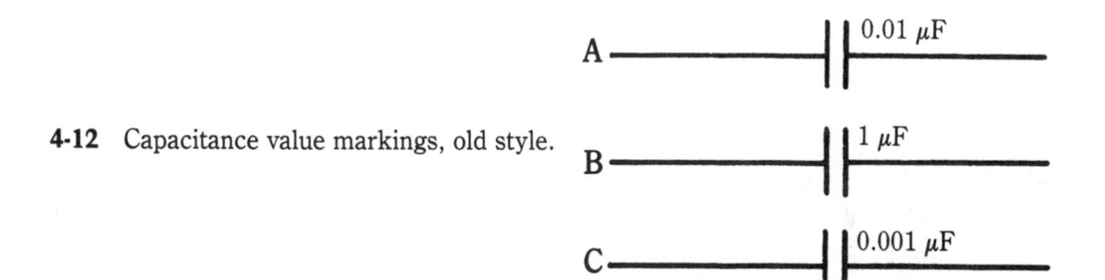

4-12 Capacitance value markings, old style.

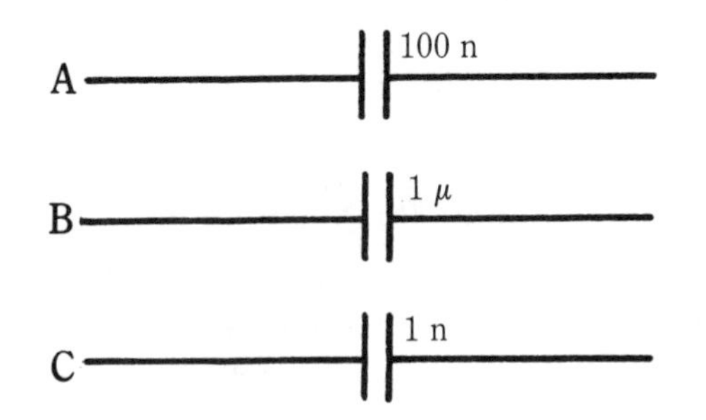

4-13 Capacitance value marking, International Standard.

eliminated. The confusion factor that 1000 k, 0.001 μF, or 1000 pF can generate in technicians of all ages and experience levels is eliminated. If a capacitor multiplier is micro-micro, pico is used. So a 47 μμ, or 0.000047 microfarad is in International Standards 47p.

The resistor problems are similar. Try to repair a unit by replacing a 3.3 k with a 33 k, because the decimal point "got lost" and you will be in trouble. Also 1 k can be written 1000 Ω and 3.3 k can be written 3300 (Fig. 4-14). Now look at Fig. 4-15.

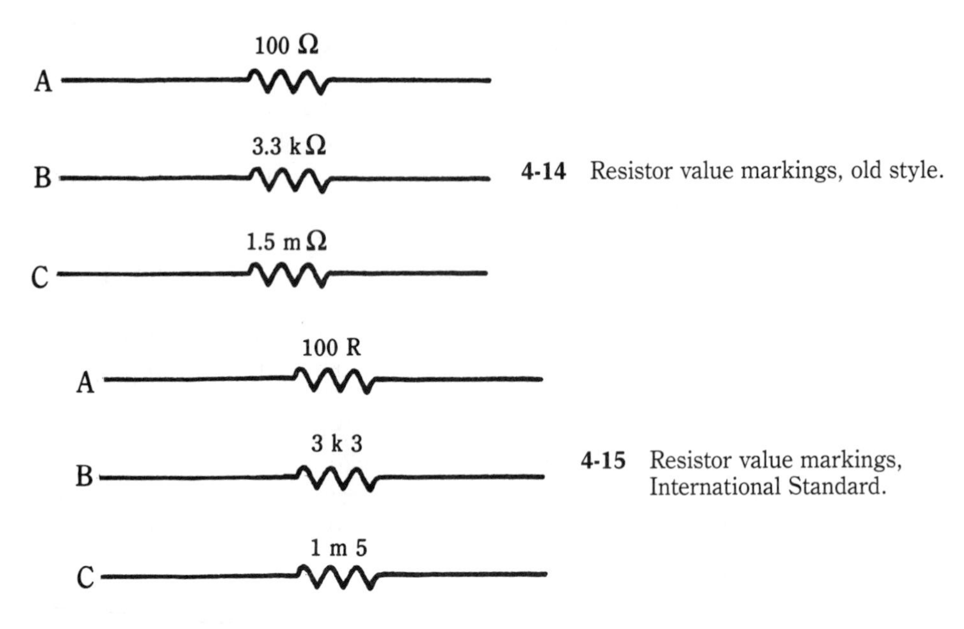

4-14 Resistor value markings, old style.

4-15 Resistor value markings, International Standard.

What problems have been solved for technicians? First is the elimination of the ohms character (Ω), which is impossible to reproduce on common typewriters and has always been a problem. A second problem solved is the illusive decimal point. The R could be dropped in a resistance that has a value of 100 R because resistance is always resistance; however, it will be needed for those resistors between 1 and 10 ohms (1.5 ohms will be 1R5, as an example).

International Standards will be used more and more in the future. Try it on some sample resistors and capacitors so that you are familiar with it. It is used in some questions in this text.

A new standard for computer and logic symbols has been developed by the International Electrotechnical Commission (IEC). In the United States the Institute of Electrical and Electronics Engineers has published the standard in its IEEE Std 91-1984. A few logic devices are shown in Fig. 4-16 and compared to the military notation. Notice the rectangular box is the basic symbol in this system. Meaning is portrayed in the qualifying symbols either in or out of the rectangle.

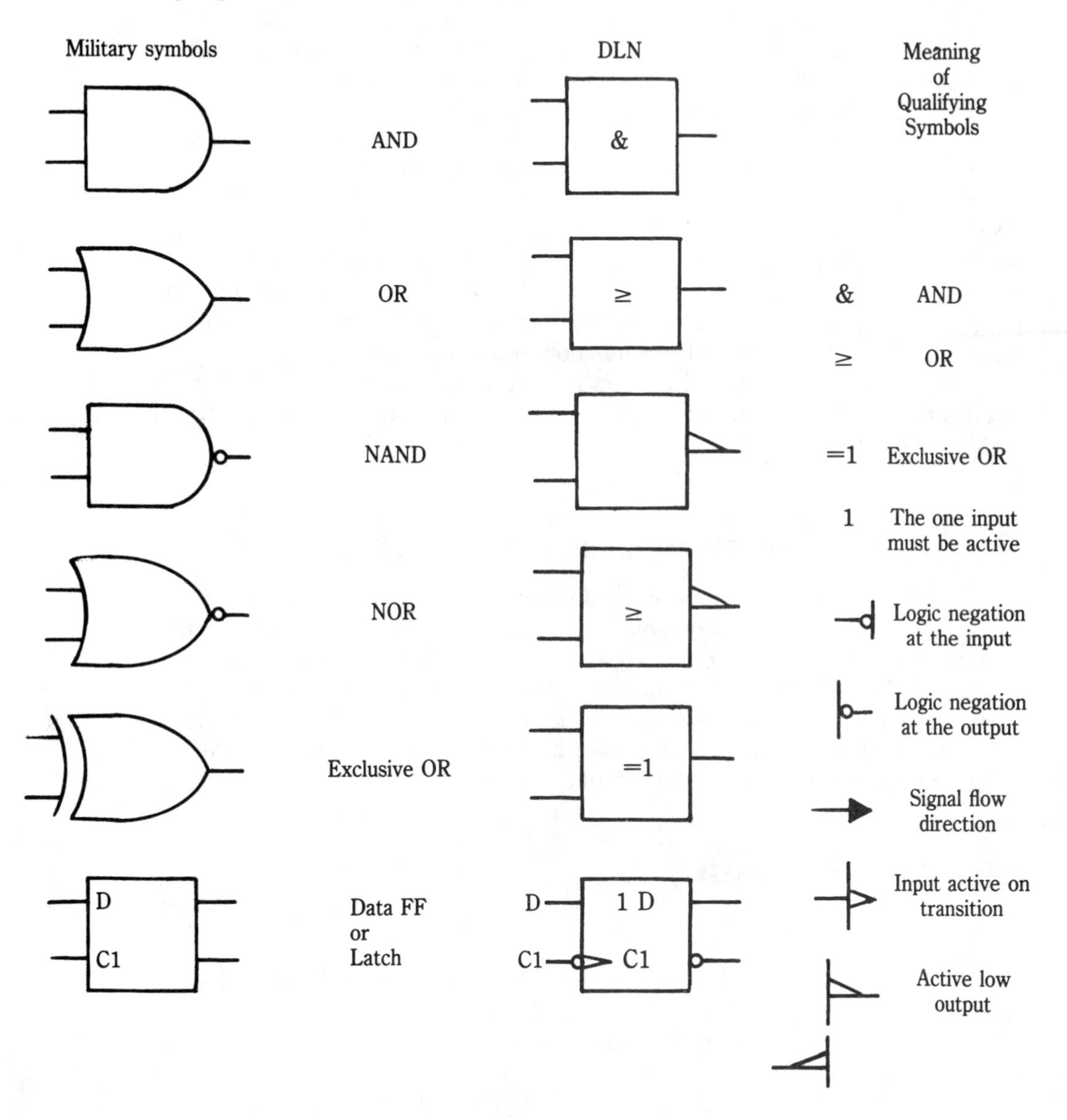

4-16 Some logic notations in DLN.

Figure 4-17 depicts a few complex logic functions found in computers.

Counters can take on various configurations. The one shown here is a binary counter, counting in this case from 0000_2 to 1111_2. Some counters count to 1001_2 or 9_{10} and then reset to 0000_2. Such a counter is called a BCD counter, a binary coded decimal counter.

Many different kinds of encoders and decoders are also available. One example is the BCD/7-segment decoder; shown here with one enable line and a lamp-test input. By looking at the patterns of "1's" shown in the truth table, for particular BCD inputs, you should be able to find the number. For example, if the BCD inputs are 0100_2, or 4_{10}, the outputs that are "1" are b, c, f, and g. If these outputs are connected to an LED 7-segment display the "4" should illuminate. A binary to 8-line decoder is also shown in the figure.

A couple of different read-only-memory symbols are also shown. The one with the four address lines (A_0 through A_3) can address 16 different locations, because $2^4 = 16$. Because there are four data lines, the ROM is said to be a 16×4 ROM. A 128×1 ROM is also shown. A RAM (random-access-memory) symbol is also shown. Notice that the data lines are shown with arrows in and out and also notice the presence of a Write/Read enable line. When this line is high, the Write function is enabled and information is stored from the data lines to the address selected by the address lines. When low, a Read function would be enabled.

Manufacturers of integrated circuits make available data books that contain specifications for their computer chips. These are often referred to by technicians working in computer based products.

Common electronics components

You know the symbols and usage of resistors, capacitors, and inductors. They are the most common electronic components. We have touched on transformers and listed the logic gates used in computerized equipment. There are many other components in addition to these that you need to become familiar with. Most are discussed in later chapters; here are many of those you will work with (Fig. 4-18):

If any of these are unfamiliar to you, be sure you concentrate on each and understand its usage as you proceed through this book.

Additional reading

Cirovic, Michael: *Integrated Circuits, A User's Handbook*, Reston Publishing Co., 1977.

Douglas-Young, John,: *Technicians Guide to Microelectronics*, Park Publishing Co., 1978.

Faber, Rodney B. and Russell L. Heiserman: *Introduction to Electron Devices*, McGraw-Hill, 1968.

Hughes, Fredrick W.: *Workbench Guide to Practical Solid State Electronics*, Parker Publishing Co., 1979.

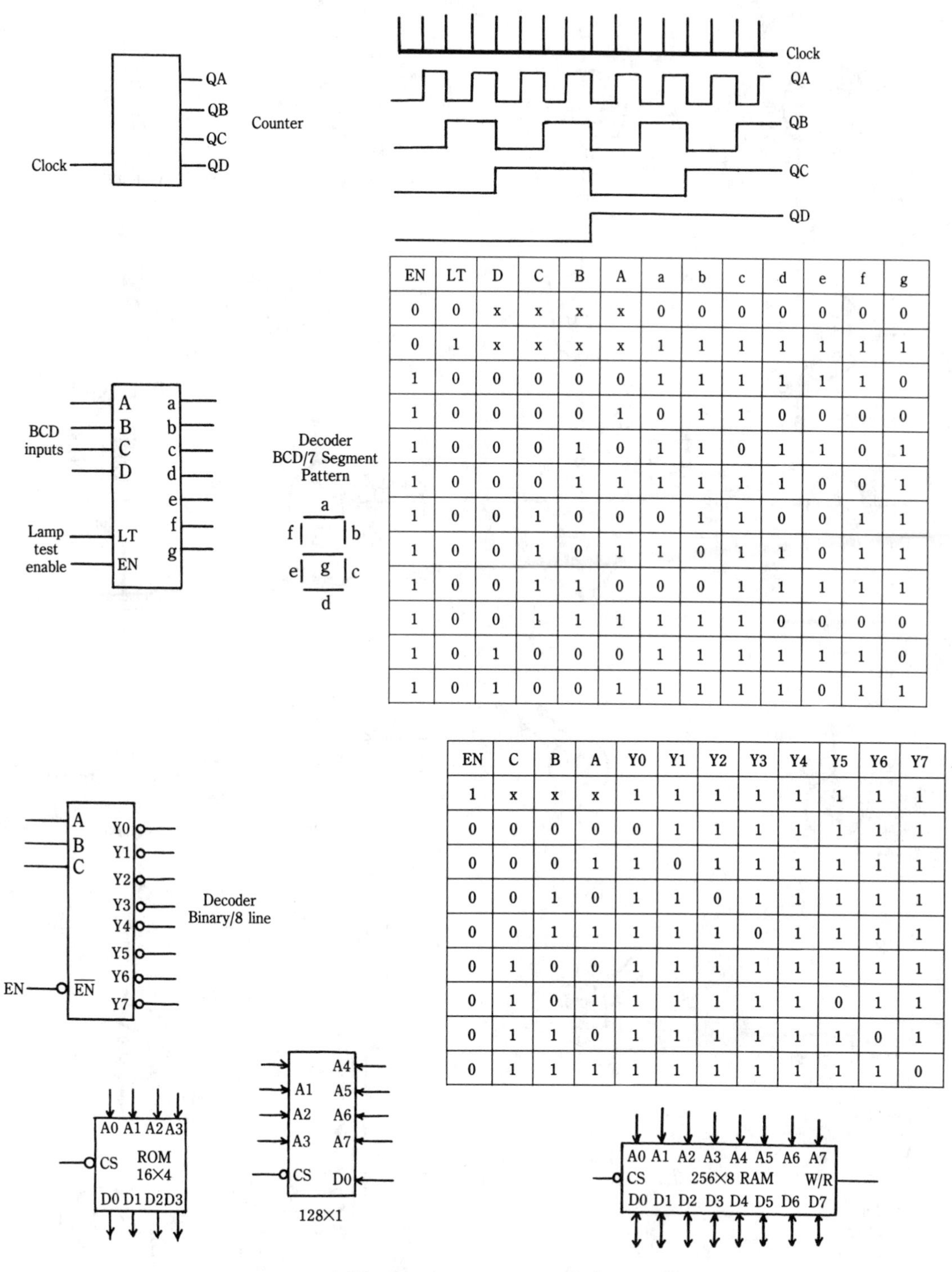

EN	LT	D	C	B	A	a	b	c	d	e	f	g
0	0	x	x	x	x	0	0	0	0	0	0	0
0	1	x	x	x	x	1	1	1	1	1	1	1
1	0	0	0	0	0	1	1	1	1	1	1	0
1	0	0	0	0	1	0	1	1	0	0	0	0
1	0	0	0	1	0	1	1	0	1	1	0	1
1	0	0	0	1	1	1	1	1	1	0	0	1
1	0	0	1	0	0	0	1	1	0	0	1	1
1	0	0	1	0	1	1	0	1	1	0	1	1
1	0	0	1	1	0	0	0	1	1	1	1	1
1	0	0	1	1	1	1	1	1	0	0	0	0
1	0	1	0	0	0	1	1	1	1	1	1	0
1	0	1	0	0	1	1	1	1	1	0	1	1

EN	C	B	A	Y0	Y1	Y2	Y3	Y4	Y5	Y6	Y7
1	x	x	x	1	1	1	1	1	1	1	1
0	0	0	0	0	1	1	1	1	1	1	1
0	0	0	1	1	0	1	1	1	1	1	1
0	0	1	0	1	1	0	1	1	1	1	1
0	0	1	1	1	1	1	0	1	1	1	1
0	1	0	0	1	1	1	1	1	1	1	1
0	1	0	1	1	1	1	1	1	0	1	1
0	1	1	0	1	1	1	1	1	1	0	1
0	1	1	1	1	1	1	1	1	1	1	0

4-17 Some computer symbols.

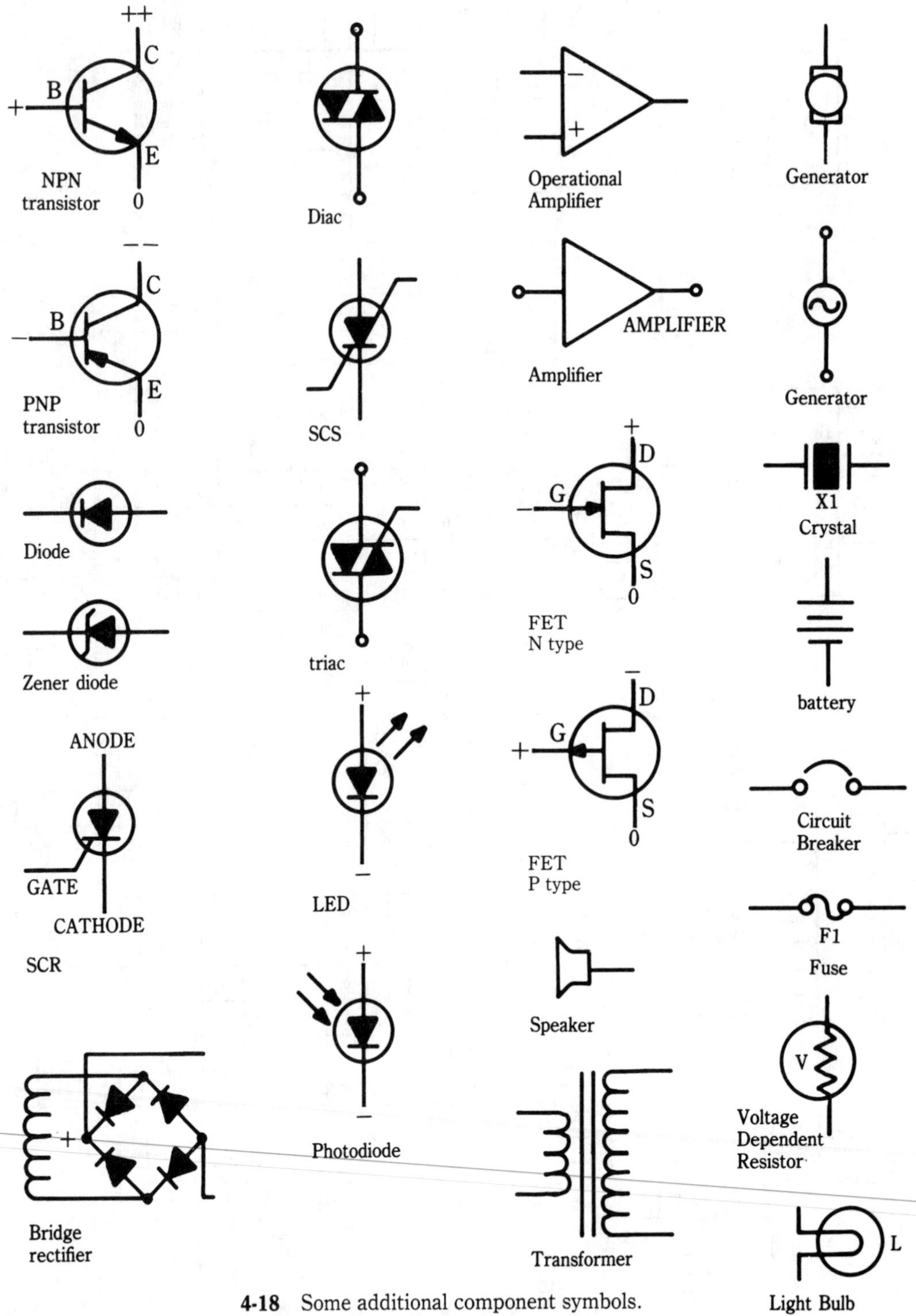

4-18 Some additional component symbols.

Jones, Thomas H.: *Electronic Components Handbook*, Reston Publishing Co., 1978.

Turner, Rufus P.: *Solid State Components*, Howard W. Sams, 1974.

Warring, R.H.: *Electronic Components Handbook for Circuit Designers*, TAB Books, 1983.

5
Series and parallel circuits

THIS CHAPTER PROVIDES A REVIEW OF THE CONCEPTS OF SERIES AND parallel circuits. As before, the quiz at the beginning covers circuit concepts important to these subjects. If you have trouble with any of these questions, review material is presented after the quiz.

Quiz

Q1. The equivalent resistance of these three resistors (Fig. 5-1) is:

a. 50 ohms
b. 15 ohms
c. nearly 5 ohms
d. None of these.

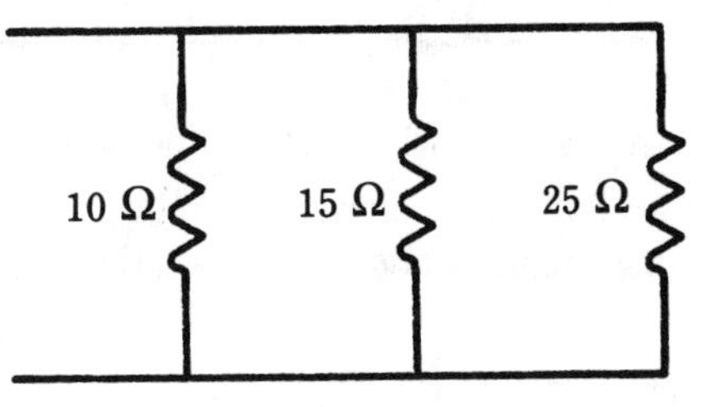

5-1 Parallel resistor circuit for Q1.

Q2. In Fig. 5-2 the resistance R equals

a. 15 kilohms
b. 50 kilohms
c. 150 kilohms
d. 35 kilohms

Q3. The equivalent capacitance of this group of capacitors (Fig. 5-3) is:

a. 125 picofarads
b. 1250 picofarads
c. 750 picofarad
d. 5 microfarads

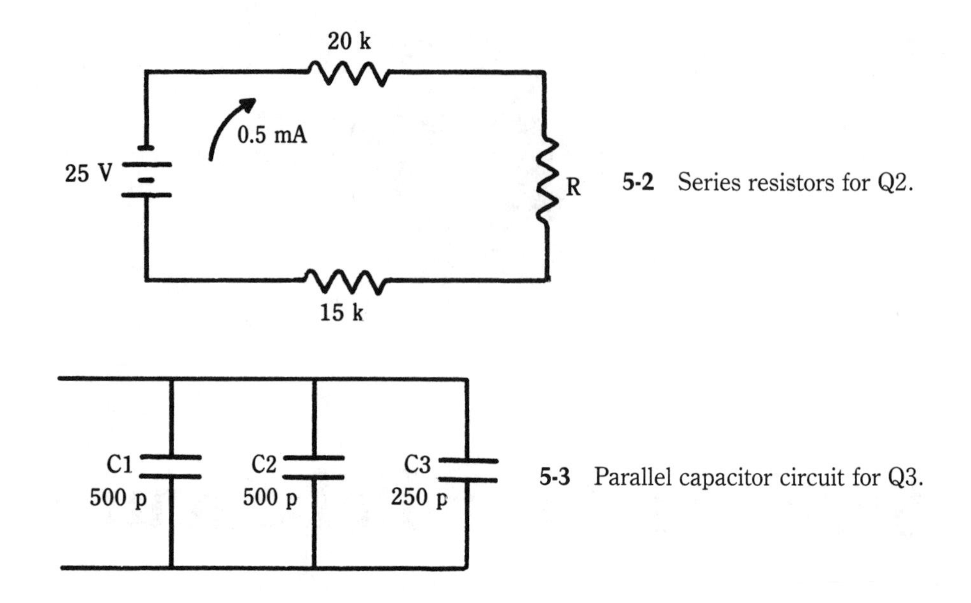

5-2 Series resistors for Q2.

5-3 Parallel capacitor circuit for Q3.

Q4. If, at some frequency, the capacitive reactances of C1, C2, and C3 in Q3 are as shown in Fig. 5-4, then the equivalent capacitive reactance is:

a. 350 ohms
b. 100 ohms
c. 400 ohms
d. 40 ohms

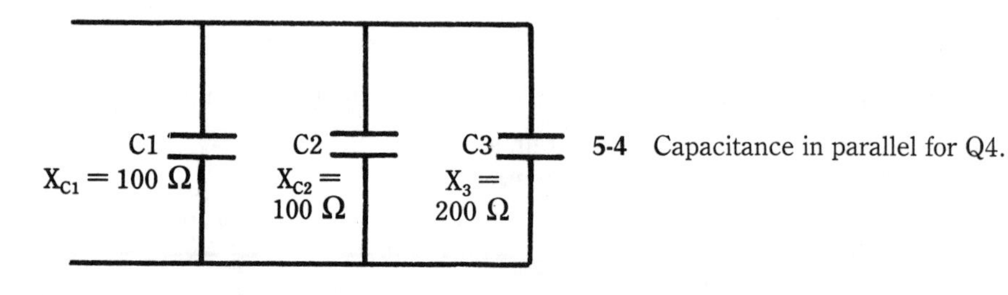

5-4 Capacitance in parallel for Q4.

Q5. In Fig. 5-5, what capacitance does the generator see?

a. 0.055 microfarad
b. 5 nanofarads
c. 0.5 microfarad
d. None of these.

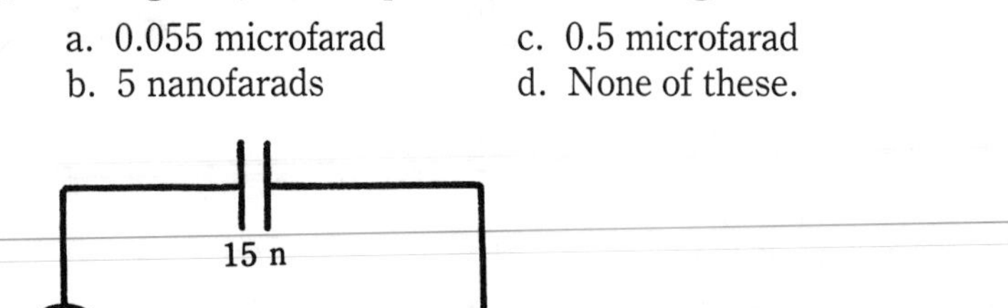

5-5 Series capacitance for problem in Q5.

Q6. In this circuit (Fig. 5-6) the voltage across C2 will be:

a. 0 Vdc c. 20 Vdc
b. 120 Vdc d. 100 Vdc

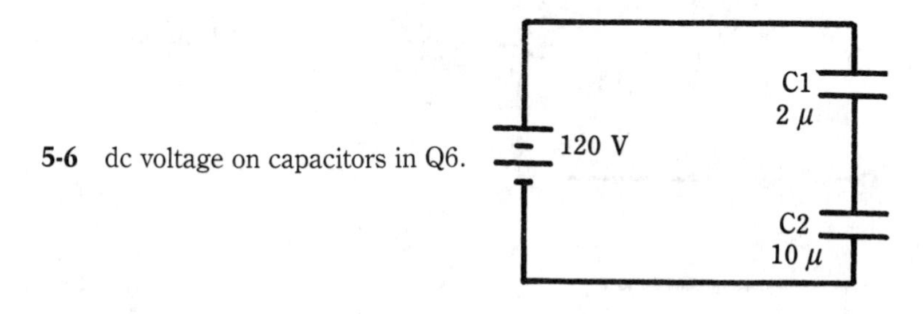

5-6 dc voltage on capacitors in Q6.

Q7. The formula for finding inductance of three parallel inductances is:

a. $L_T = L1 + L2 + L3$

b. $L_T = \frac{1}{L_1} + \frac{1}{L_2} + \frac{1}{L_3}$

c. $L_T = \frac{1}{\frac{1}{L1} + \frac{1}{L2} + \frac{1}{L3}}$

d. None of these.

Q8. In the circuit of Fig. 5-7, find the ac voltage drop across $L2$:

a. 2.5 Vac c. 0 volt at this frequency
b. 7.5 Vac d. 10 volts at this frequency

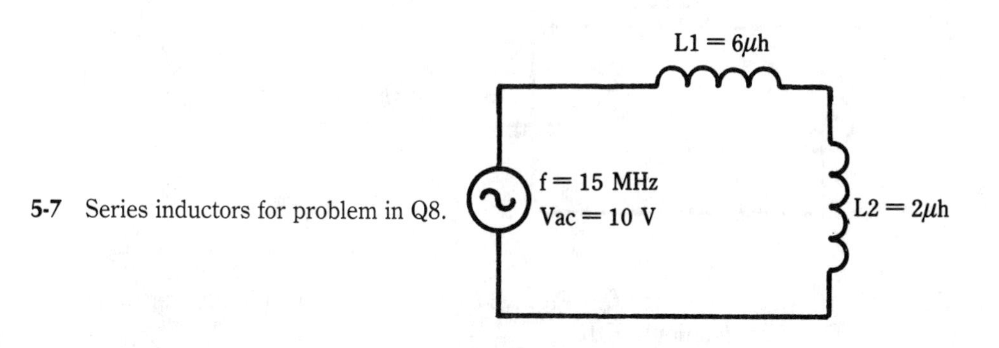

5-7 Series inductors for problem in Q8.

Q9. In the circuit of Fig. 5-8, the combination of impedances (Zin) is:

a. 40 ohms c. 10 ohms
b. 20 ohms d. 28.3 ohms

Q10. Find the impedance of the parallel combination of R and C in the circuit of Fig. 5-9:

a. 240 ohms c. 700 ohms
b. 350 ohms d. 500 ohms

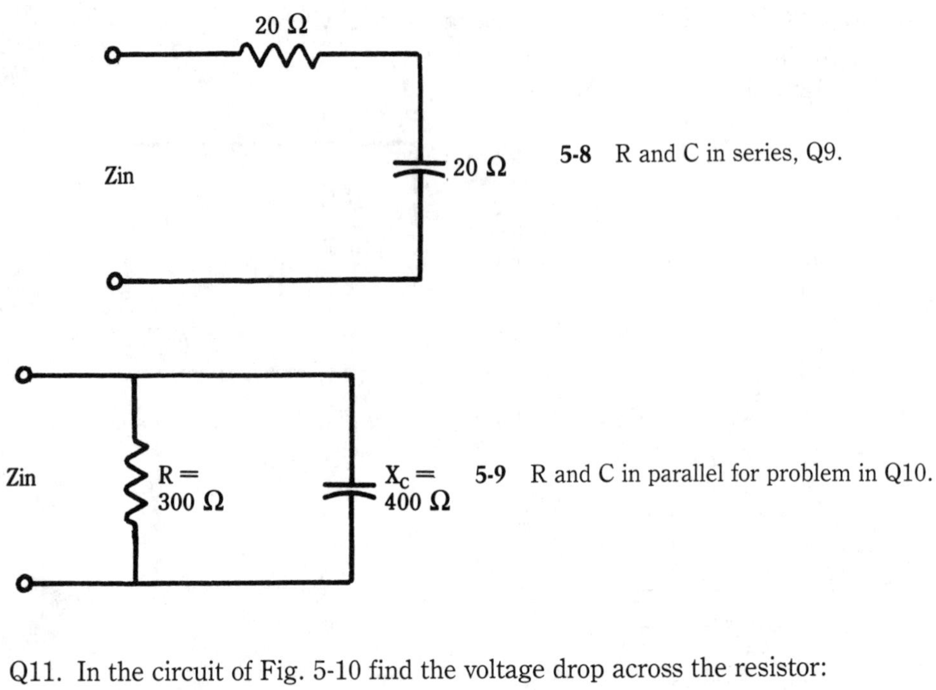

5-8 R and C in series, Q9.

5-9 R and C in parallel for problem in Q10.

Q11. In the circuit of Fig. 5-10 find the voltage drop across the resistor:

a. 35 volts
b. 30 volts
c. 10 volts
d. 22 volts

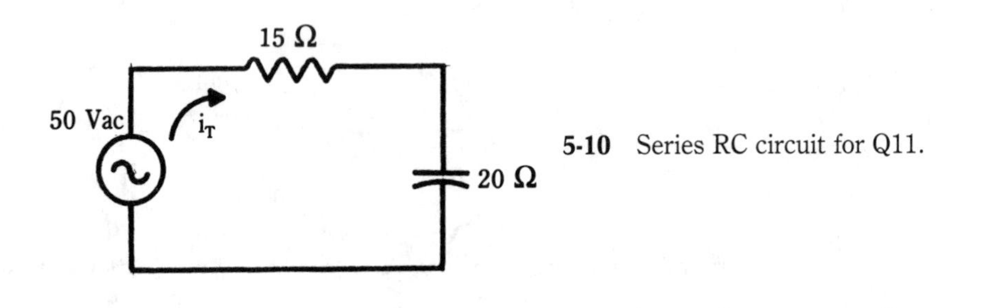

5-10 Series RC circuit for Q11.

Q12. The output of the circuit in Fig. 5-11 will respond to an input step voltage in the following fashion:

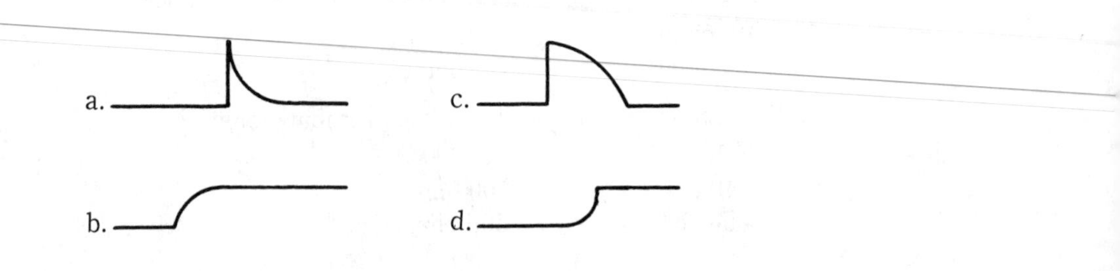

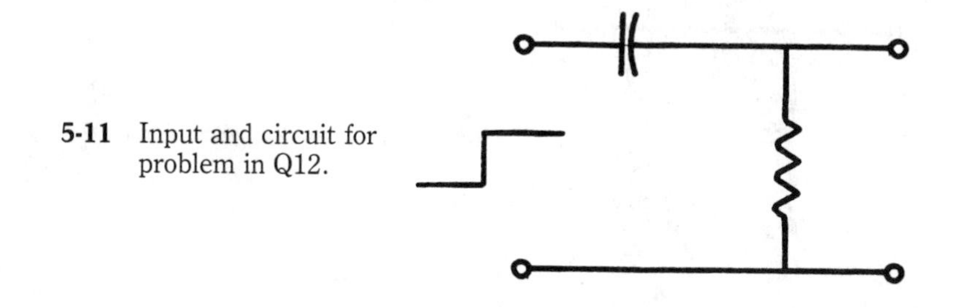

5-11 Input and circuit for problem in Q12.

Q13. In the circuit of Fig. 5-12, the output voltage will reduce to 7.4 volts after a rise of 20 volts at the occurrence of the input step voltage. It will take how much time to do this?

a. 1 millisecond
b. 1 microsecond
c. 1 second
d. None of these.

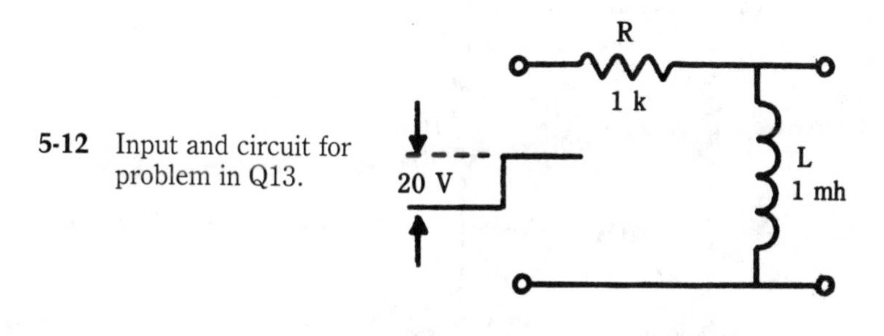

5-12 Input and circuit for problem in Q13.

Q14. In a parallel LC tank circuit like Fig. 5-13 if the frequency of the applied voltage is less or lower than the resonant frequency the circuit will appear:

a. capacitive
b. resistive
c. inductive
d. None of these.

Q15. If $X_L = 30$ ohms and $X_C = 40$ ohms in the circuit of Fig. 5-13, what combined impedance will the generator see?

a. 120 ohms
b. 70 ohms
c. 10 ohms
d. None of these.

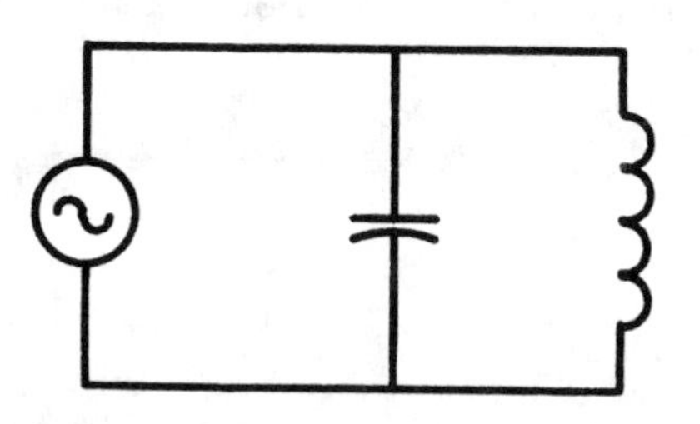

5-13 Tank circuit for Q14 and Q15.

Q16. A parallel resonant circuit connected as shown in Fig. 5-14 would be:

a. a trap at f_0
b. a low-pass filter
c. a high-pass filter
d. a bandpass filter

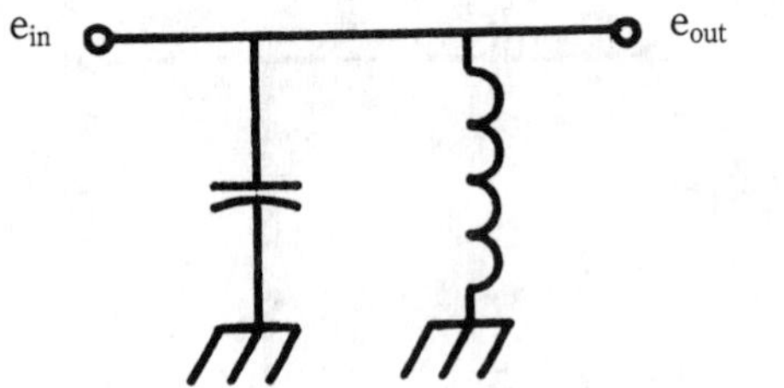

5-14 Resonant circuit for Q16.

Q17. In the circuit of Fig. 5-15, what is the voltage at A with respect to B?

a. −8 volts
b. +8 volts
c. +10 volts
d. +4 volts

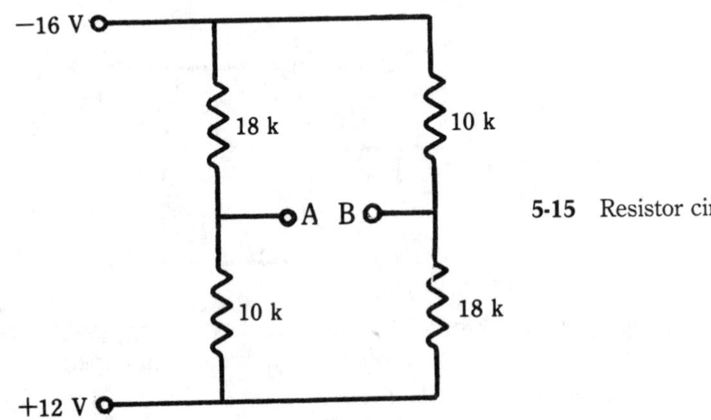

5-15 Resistor circuit problem, Q17.

Q18. In the circuit of Fig. 5-16 R1 is a brown, green, red 20-percent resistor but its actual resistance is:

a. 1500
b. 1.67 k
c. out of tolerance
d. Both b and c are correct.

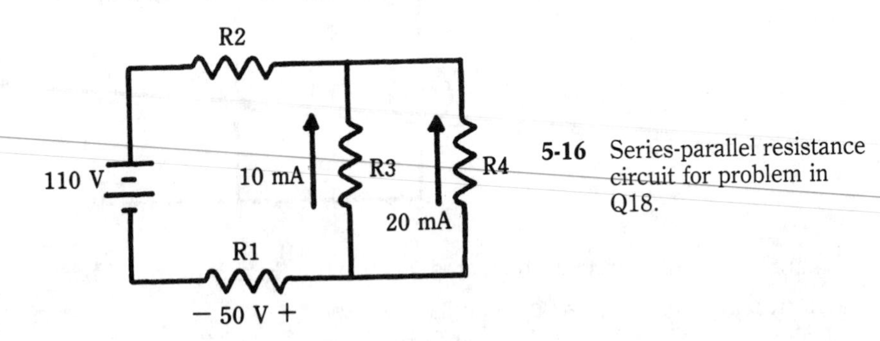

5-16 Series-parallel resistance circuit for problem in Q18.

Parallel circuits (resistive)

The first question in the quiz (Q1) deals with resistors in parallel. Parallel resistances each draw current from the external circuit; therefore, the external circuit "sees" a total of the three currents. This total is larger than any individual current. If this current were supplied to an equivalent resistance, the equivalent resistance would be smaller than any one of the three. Hence the correct answer to Q1 is C, nearly 5 ohms.

Another way to solve this problem is to assume some convenient voltage and then calculate individual currents. Summing the individual currents and dividing into the assumed voltage will result in the equivalent resistance. For example, assume for question one a voltage of 150 volts. The three currents would then be:

$$\frac{150\ \text{V}}{10} = 15\ \text{A}$$

$$\frac{150\ \text{V}}{15} = 10\ \text{A}$$

$$\frac{150\ \text{V}}{25} = 6\ \text{A}$$

The total current is then 15 A + 10 A + 6 A = 31 A; and the equivalent resistance is 150 V/31 A which equals a little less than 5 ohms.

Series circuits (resistive)

In series circuits, each circuit element has the entire current flowing through it. The voltage drops around a closed loop must equal zero. In Q2 because we are dealing with a circuit in which all the current goes through each element, it is a series circuit. For this problem we are already given the value of the current. Because we also know the source voltage, we can easily calculate the resistance the source "sees." Let's call it R_T.

Then $R_T = 25\ \text{V} \div 0.5\ \text{mA} = 50\ \text{k}\Omega$
and because R_T also $= R + 20\ \text{k}\Omega + 15\ \text{k}\Omega$
then $R + 20\ \text{k}\Omega + 15\ \text{k}\Omega = 50\ \text{k}\Omega$
$R = 50\ \text{k}\Omega - 35\ \text{k}\Omega$
$= 15\ \text{k}\Omega$

Therefore, the correct answer to Q2 is answer a. 15 k.

Just for review let's use this same circuit to calculate the voltage drops around this loop to make sure they total zero. Because each voltage drop across a resistor is the current (0.5 mA) multiplied by the resistance then the three resistance voltage drops are:

$(0.5\ \text{mA})(20\ \text{k}\Omega) = 10\ \text{V}$
$(0.5\ \text{mA})(15\ \text{k}\Omega) = 7.5\ \text{V}$
and $(0.5\ \text{mA})(15\ \text{k}\Omega) = 7.5\ \text{V}$

Now, remembering that in going around a loop it is conventional to assign minuses to voltage drops and pluses to voltage rises we have:

$$+25\text{ V} - 10\text{ V} - 7.5\text{ V} - 7.5\text{ V} = 0$$

Capacitors in parallel

The capacitance of capacitors in parallel adds; hence, a 1000-picofarad capacitor in parallel with a 500-picofarad capacitor would be 1500 picofarads. Or, as in Q3, a 500-picofarad capacitor in parallel with a 500-picofarad capacitor and also in parallel with a 250-picofarad capacitor would be:

$$500\text{ pF} + 500\text{ pF} + 250\text{ pF} = 1250\text{ pF}$$

But, the capacitive reactances of capacitors in parallel act like the resistances of resistors in parallel. That is:

$$\frac{1}{X_{CT}} = \frac{1}{X_{C1}} + \frac{1}{X_{C2}} + \frac{1}{X_{C3}}$$

$$\text{or } X_{CT} = \frac{1}{\dfrac{1}{X_{C1}} + \dfrac{1}{X_{C2}} + \dfrac{1}{X_{C3}}}$$

Therefore, to find the equivalent reactance in Q4:

$$X_{CT} = \frac{1}{\dfrac{1}{100} + \dfrac{1}{100} + \dfrac{1}{200}}$$

$$= \frac{1}{\dfrac{2+2+1}{200}}$$

$$= \frac{200}{2+2+1}$$

$$= 40\ \Omega$$

This concept is difficult to remember, so it might help to approach it in this way. Because $C_T = C_1 + C_2 + C_3$ (three capacitors in parallel) and because

$$X_c = \frac{1}{2\pi fC}\text{: or } C = \frac{1}{2\pi fX_c}$$

$$\text{then } \frac{1}{2\pi fX_{CT}} = \frac{1}{2\pi fX_{C1}} + \frac{1}{2\pi fX_{C2}} + \frac{1}{2\pi fX_{C3}}$$

then multiplying both sides of this equation by $2\pi f$ gives

$$\frac{1}{X_{CT}} = \frac{1}{X_{C1}} + \frac{1}{X_{C2}} + \frac{1}{X_{C3}}$$

Another memory aid would be to replace the capacitors in the circuit with their capacitive reactances, as in Fig. 5-17. You can then work the problem like a resistor problem.

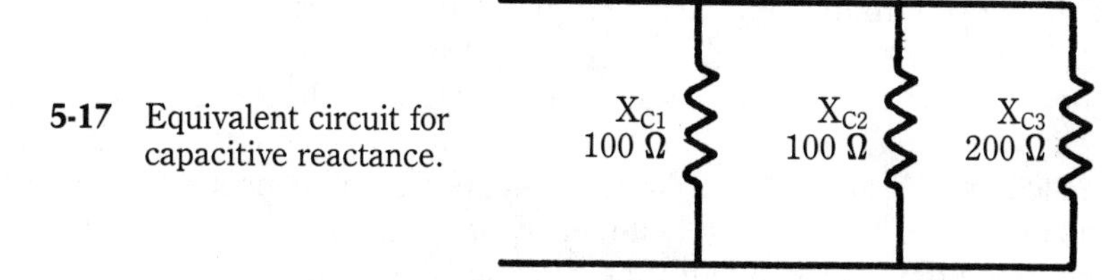

5-17 Equivalent circuit for capacitive reactance.

Capacitors in series

The capacitance of capacitors in series is combined in similar fashion to resistances in parallel, that is:

$$\frac{1}{C_T}=\frac{1}{C_1}+\frac{1}{C_2}+\frac{1}{C_3}$$

Then to solve Q5:

$$\begin{aligned}\frac{1}{C_T}&=\frac{1}{15\text{ nF}}+\frac{1}{30\text{ nF}}+\frac{1}{10\text{ nF}}\\&=\frac{2}{30\text{ nF}}+\frac{1}{30\text{ nF}}+\frac{3}{30\text{ nF}}\\&=\frac{2+1+3}{30\text{ nF}}\\&=\frac{6}{30\text{ nF}}\end{aligned}$$

Taking the reciprocal of both sides gives:

$$\begin{aligned}C_T&=\frac{30\text{ nF}}{6}\\&=5\text{ nF}\end{aligned}$$

Suppose the circuit was the same but the input frequency and current were given. The voltage drops around the circuit could be calculated by finding capacitive reactances. For example: To find the voltage drop across C2 in Fig. 5-18:

Because we know $C_2=30$ nF

$$\begin{aligned}X_{C2}&=\frac{1}{2\pi fC_2}\\&=\frac{1}{2\pi(2\times10^6)\,(30\times10^{-9})}\\&=26.7\text{ ohms}\end{aligned}$$

$$V_{C2} = IX_{C2}$$
$$= 2 \text{ mA } (26.7)$$
$$= 53.4 \text{ mV}$$

If this is giving you trouble, practice by finding voltages for C_1 and C_3 and for C_T. If calculations are correct $V_{CT} = V_{C1} + V_{C2} + V_{C3}$. Do they?

It should be no surprise to notice that the voltage drop is largest across the largest capacitive reactance. But also notice that the *smallest* capacitor has the largest reactance. Also when a technician wants a small capacitor, reference is to the capacitance. A 100-picofarad is much *smaller* than a one-microfarad capacitor, even though the physical size might be the opposite.

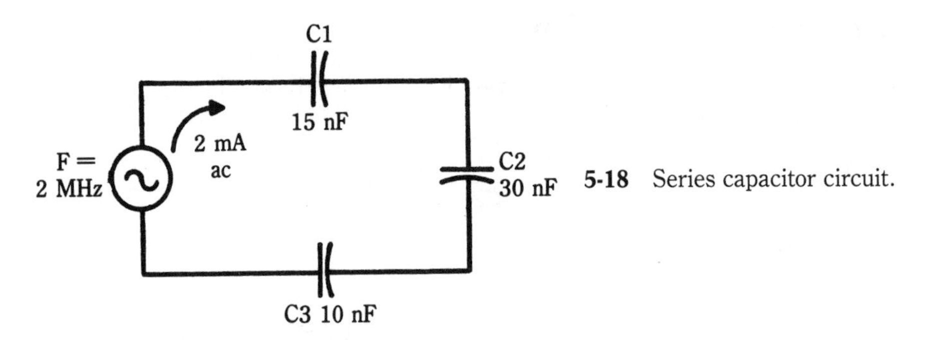

5-18 Series capacitor circuit.

Voltage across capacitors in a dc circuit also distribute in an inverse relationship. That is the smaller capacitor will have the largest voltage drop. This is a tough one to comprehend, partly because some of the things studied about capacitors seem to contradict this. We all learned that "capacitors block dc." With that in mind, looking at the circuit in Q6 might lead to the conclusion that because C1 "can't pass dc" then the voltage across C2 ought to be zero volts. But if that were true then the voltage across C1 ought likewise to be zero, and then Kirchhoff's voltage law wouldn't hold up. In actuality, in this circuit each capacitor charges initially as the electrons flow around the circuit when it is first connected up. Each has the same charge of electrons (careful here, charge is not the same as voltage), but because the capacitors have different capacitances—different sizes—the electrons per square inch are different. That is, the voltage across one is different than the voltage across the other. The relationship is shown mathematically as:

$$Q = CE$$

where Q = charge

Apply this to the circuit in Q6

$$Q1 = Q2$$

and $Q1 = C1E_{C1}$

where E_{C1} = the voltage across C1 and $Q2 = C2E_{C2}$, which gives $C1E_{C1} = C2E_{C2}$ or

$$\frac{E_{C1}}{E_{C2}} = \frac{C_2}{C_1}$$

This equation says that the voltage drops are inversely proportional to the capacitances. And Kirchhoff's laws tell us that:

$E_{C1} + E_{C2} = 120$ volts in this circuit. Substituting C1 and C2 values in the ratio equation gives:

$$\frac{E_{C1}}{E_{C2}} = \frac{10\ \mu F}{2\ \mu F}$$

Hence: $E_{C1} = E_{C2} \bullet 5$

Substituting this in Kirchhoff's equation and solving for E_{C2} gives:

$$5E_{C2} + E_{C2} = 120$$
$$E_{C2}(5 + 1) = 120$$
$$E_{C2} = 20\ V$$

Occasionally technicians need to substitute or try a capacitor in a circuit and they find that to obtain the capacitance they desire they have to put two capacitors in series. The above distribution of voltages should be kept in mind when selecting working voltages if dc voltage is involved in the circuitry.

Let's look at a circuit just to examine a hypothetical use. Suppose that in the circuit of Fig. 5-19 you suspect the bypass capacitor C2 and don't have the exact value, but you do have a 200 μF@10 WV and a 300 μF@ 50 WV. If you make the substitution, the emitter circuit will look like the circuit in Fig. 5-20.

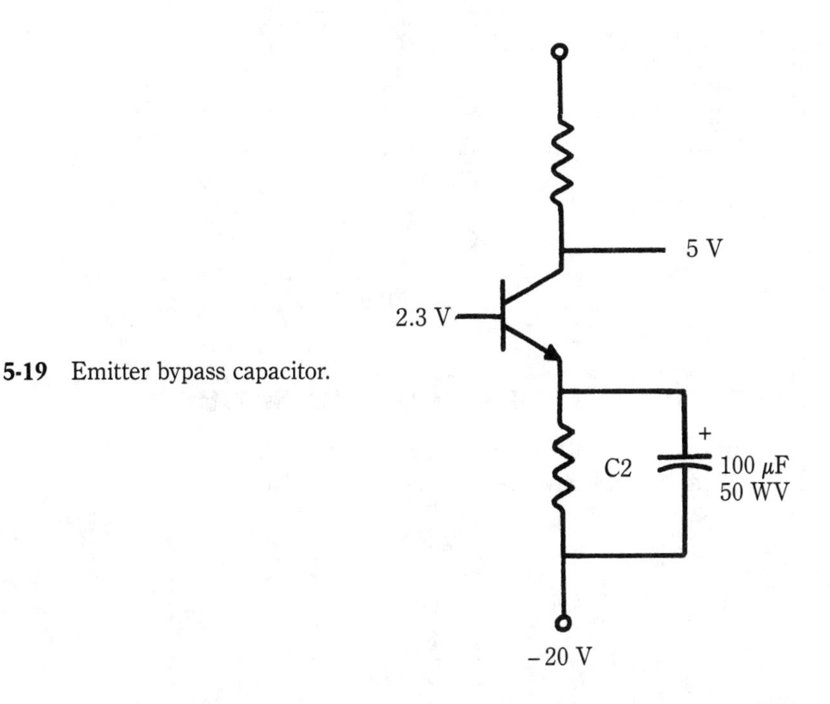

5-19 Emitter bypass capacitor.

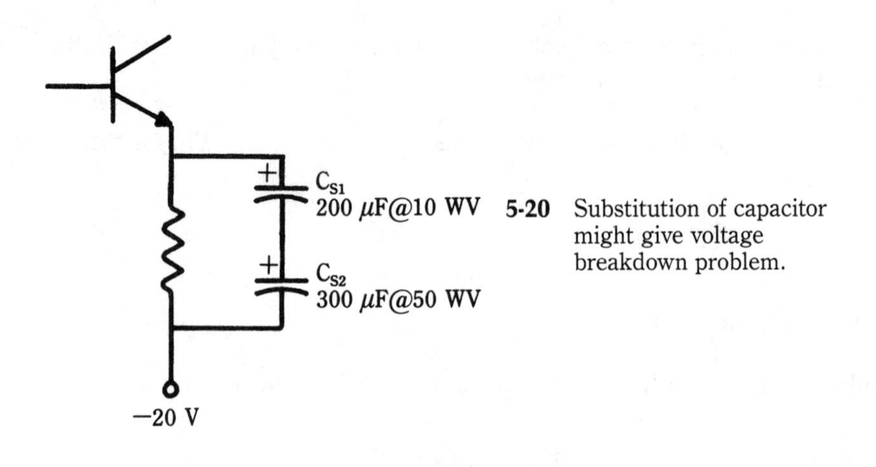

5-20 Substitution of capacitor might give voltage breakdown problem.

Will C_{S1} and C_{S2} be OK? You can expect in excess of 20 volts drop across the emitter resistor. This voltage will divide 2/5 across C_{S2} and 3/5 across C_{S1}, or C_{S2} will drop in excess of 2/5 (20) = 8 volts and C_{S1} will drop in excess of 3/5(20) = 12 volts.

Chances are that C_{S1} will not last long because its rated working voltage is less than the dc drop it will encounter used in this way. The capacitance will be close to the right value as calculated here (120 μF).

$$\frac{1}{C_T} = \frac{1}{C_{S1}} + \frac{1}{C_{S2}}$$

$$\frac{1}{C_T} = \frac{1}{200} + \frac{1}{300}$$

$$\frac{1}{C_T} = \frac{3+2}{600\ \mu\text{F}}$$

$$\frac{1}{C_T} = \frac{5}{600\ \mu\text{F}}$$

$$C_T = \frac{600\ \mu\text{F}}{5}$$

$$= 120\ \mu\text{F}$$

Inductances in parallel and series

Inductors in parallel add like resistances in parallel. That is, each inductance in parallel reduces the effective parallel inductance. Hence:

$$\frac{1}{L_T} = \frac{1}{L_1} + \frac{1}{L_2} + \frac{1}{L_3}$$

$$\text{or}\quad L_T = \frac{1}{\frac{1}{L_1} + \frac{1}{L_2} + \frac{1}{L_3}} \qquad \text{(Answer to Q7.)}$$

There are several ways to solve Q8. One way is to use Ohm's law and the inductive reactances to find the circuit's current. Then multiply that current by the X_L (inductive reactance) of L2 to find the voltage. Another way is to do those exact steps but not solve for the numerical value until the end. If we do it this way we won't actually ever have to solve for the value of the current.

$$i_T = \frac{10\text{ V}}{X_{LT}}$$

where X_{LT} = total inductive reactance

$= X_{L1} + X_{L2}$

(X_{L1} = reactance of L1; X_{L2} = reactance of L2)

$= 2\pi f\, L_1 + 2\,\pi f\, L_2$

But the voltage across L2 is $i_T\,(X_{L2})$.

Therefore:

$$V_{L2} = i_T\,(X_{L2}) = \frac{10\text{ V}}{X_{CT}}\,(X_{L2})$$

$$= \frac{10\text{ V}\,(2\pi f L_2)}{2\pi f L_1 + 2\pi f L_2)} = \frac{10\text{ V}\,(2\pi f)\,L_2}{(2\pi f)(L_1 + L_2)}$$

$$= 10\text{ V}\left(\frac{L_2}{L_1 + L_2}\right)$$

now substituting values:

$$V_{L2} = 10\text{ V}\left(\frac{2\text{ h}}{6\text{ h} + 2\text{ h}}\right)$$

$$= 10\text{ V } 2/8$$

$$= 2.5\text{ Vac}$$

To answer a question such as Q8, the preceding math will give the correct answer, a, or 2.5 volts ac. But why not merely analyze the circuit for an easier solution? All of the ac voltage must be dropped across the two series inductors. The larger percentage of the 10 volts ac will be dropped across the larger inductor. Because the larger inductor appears to have about 75 percent of the total inductive reactance, then about 75 percent of the total voltage will be dropped across L1. Because 100 percent and 0 percent aren't legitimate answers, you can choose 7.5 volts ac as the drop across L1, or 2.5 volts. If you correctly choose 7.5 volts, then the remaining 2.5 volts is dropped across L2.

RL and RC circuits

Series and parallel RC and RL circuits are a little more complicated because the current and voltage might not be in phase with each other. This necessitates addition or combinations by vectors. For example, in the series RC circuit of Q9, the voltages are 90° out of phase. If we use the voltage across the resistor as our reference and remember that current leads voltage in a capacitor, then we can make a vector drawing to represent the relationships (Fig. 5-21).

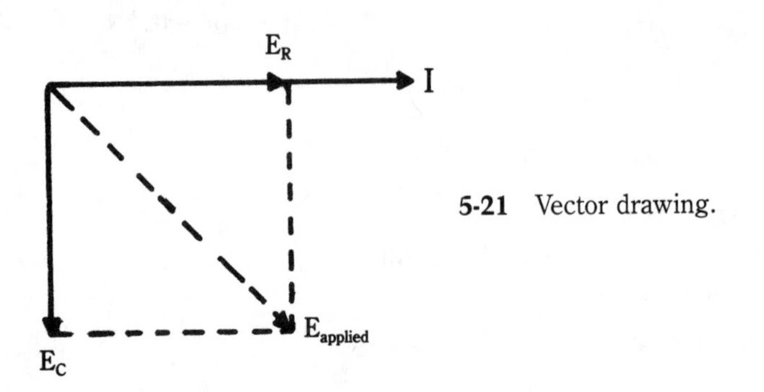

5-21 Vector drawing.

The reactance relationships have the same phase relationship (Fig. 5-22).

Because $X_c = 20$ ohms in this problem and because $R = 20$ ohms, we could estimate X_{in} by constructing our vector diagram carefully on a graph where the lengths are proportional to the ohmic values as in Fig. 5-23. Even a rough graph like this shows that the length (value) of Z_{in} is between 25 and 30 ohms. Thus, if there is to be a correct answer in Q9, it must be d, 28.3 ohms.

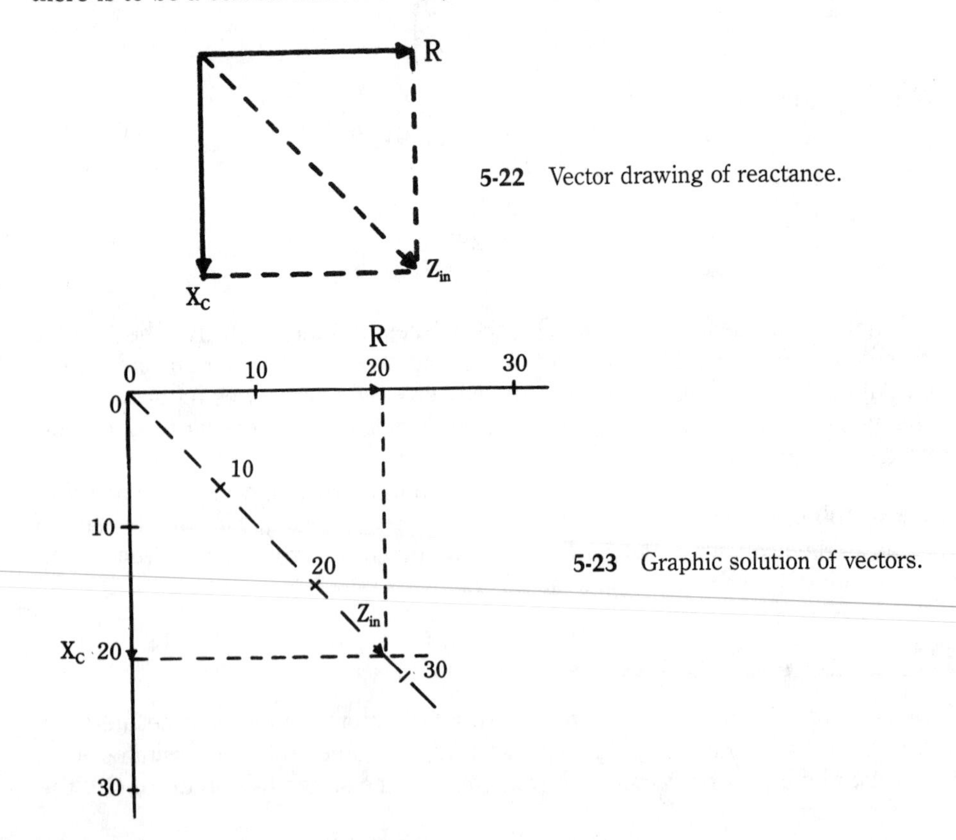

5-22 Vector drawing of reactance.

5-23 Graphic solution of vectors.

The problem also can be solved using the Pythagorean theorem or by the trigonometric functions. We'll solve the next problem by the Pythagorean theorem and the next with trigonometry.

In Q10, the vector drawing shown in Fig. 5-24 could be constructed. Notice that the voltage is used as reference because E is the same on the parallel components; but again, I_C leads the voltage in the capacitor and I_R is in phase with the voltage in the resistor. Again, the applied voltage will lag the resultant circuit current (Fig. 5-25).

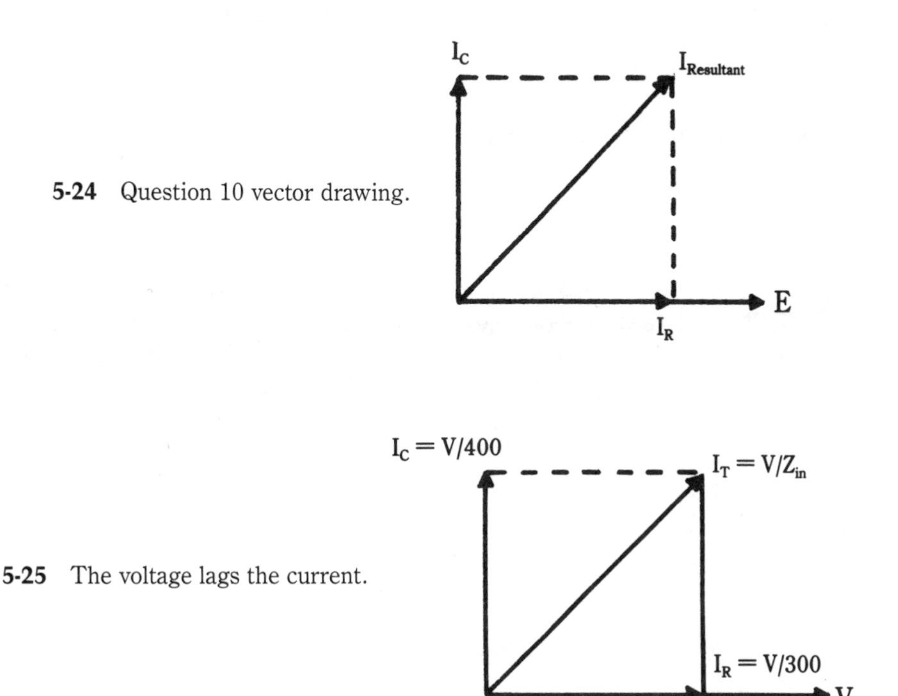

5-24 Question 10 vector drawing.

5-25 The voltage lags the current.

Pythagoras discovered that in right triangles the hypotenuse squared was equal to the sum of the square of the two sides. The vector diagram can be arranged easily into a right triangle (Fig. 5-26).

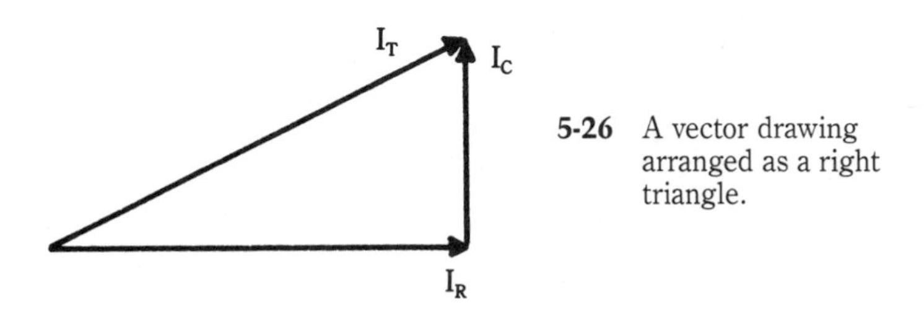

5-26 A vector drawing arranged as a right triangle.

Then it follows that the hypotenuse:

$$\left(\frac{V}{Z_{in}}\right)^2 = \left(\frac{V}{X_C}\right)^2 + \left(\frac{V}{R}\right)^2$$

Now divide both sides of the equation by V^2.

$$\left(\frac{V}{Z_{in}}\right)^2 = \left(\frac{V}{400}\right)^2 + \left(\frac{V}{300}\right)^2$$

$$\left(\frac{1}{Z_{in}}\right)^2 = \frac{1}{160{,}000} + \frac{1}{90{,}000}$$

Putting the right side of the equation over a common denominator.

$$= \frac{90\ \text{k}\Omega + 160\ \text{k}\Omega}{(160\ k\Omega)\ (90\ \text{k}\Omega)}$$

Take the reciprocal and then the square root of both sides:

$$Z_{in} = \sqrt{\frac{(160\ \text{k}\Omega)\ (90\ \text{k}\Omega)}{250\ \text{k}\Omega}} = \sqrt{57.6\ \text{k}\Omega} = 240$$

Question 11 can be solved using either method shown above to find the circuit impedance. Then find the current by $i_T = 18 \div Z_{in}$ and then by multiplying i_T by the resistance to obtain the voltage drop across the resistor. But the impedance also can be found using trigonometric functions. Not all schools of electronics teach these principles because the previous method works quite well. We include this section for review by those technicians familiar with these techniques. Again, we construct a triangle (Fig. 5-27).

The angle $\theta = \tan^{-1} 20/15$ (read as angle whose tangent is 20/15)

$= 53°\ 8'$ (look up in table)

and $Z_{in} = 20/\sin\theta$

$= 20/(0.8)$ [look up in table]

$= 25$

then $i_T = \text{Vac}/Z_{in}$

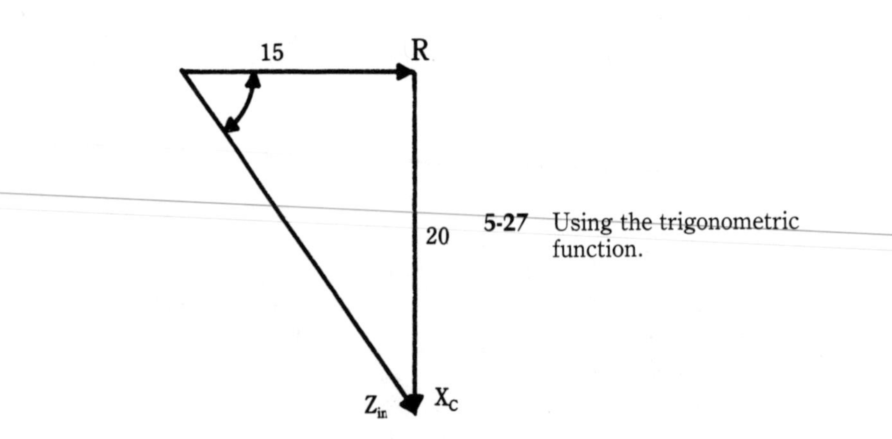

5-27 Using the trigonometric function.

$$i_T = \frac{50}{25}$$
$$= 2\text{A}$$

$$V_R = i_T R$$
$$= 2\ (15)$$
$$= 30 \text{ volts (Answer to Q11.)}$$

Any math book probably contains the trigonometric formulas. Here is a listing of a few common formulas (Fig. 5-28).

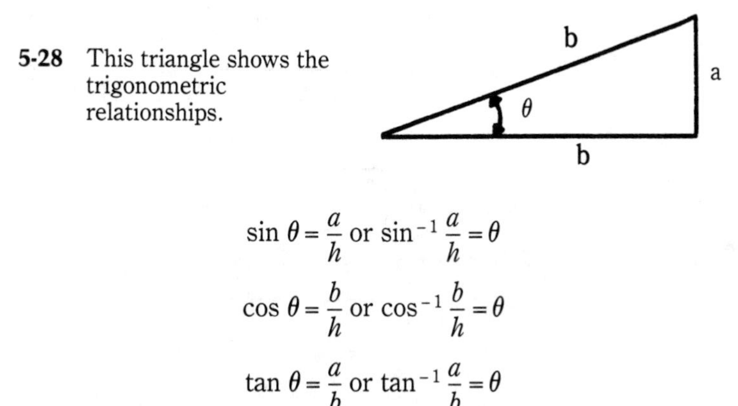

5-28 This triangle shows the trigonometric relationships.

$$\sin\theta = \frac{a}{h} \text{ or } \sin^{-1}\frac{a}{h} = \theta$$

$$\cos\theta = \frac{b}{h} \text{ or } \cos^{-1}\frac{b}{h} = \theta$$

$$\tan\theta = \frac{a}{b} \text{ or } \tan^{-1}\frac{a}{b} = \theta$$

Capacitors and resistors in series are often used in circuits with nonsinusoidal waveforms—often with square waves or pulses. Question 12 deals with this type of situation. Here we need to be aware that capacitors take time to charge up. In fact, the time it takes is proportional to the product of resistance and capacitance.

The time it takes a capacitor to charge or discharge by 63 percent of the total charge possible, is the product of R (the resistance of the charge or discharge path) and C (the capacitance of the capacitor).

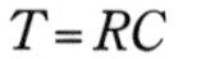

$$T = RC$$

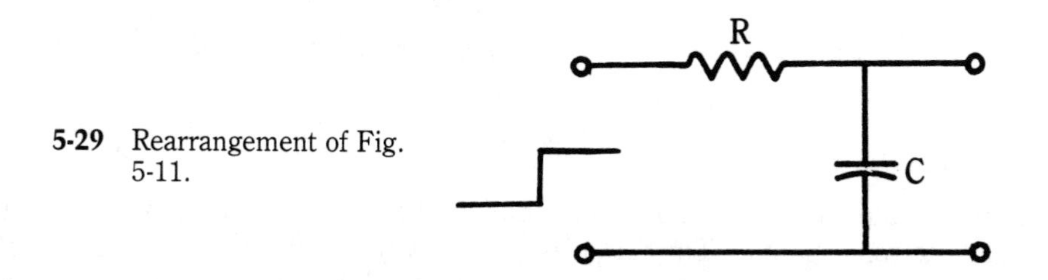

5-29 Rearrangement of Fig. 5-11.

Before tackling Q12, let's examine the circuit in Fig. 5-29. It's the same circuit but rearranged to emphasize the voltage across the capacitor. As we just said, the capacitor takes time to charge; hence, we would expect the output to start at zero and charge eventually (in at least 10 RC time constants) to the input voltage.

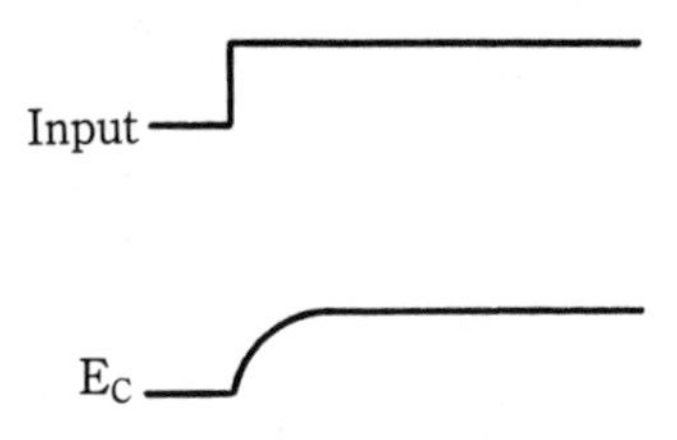

Now keeping in mind that the voltage around the series circuit must add up to zero at all times, what must the voltage be across the resistor?

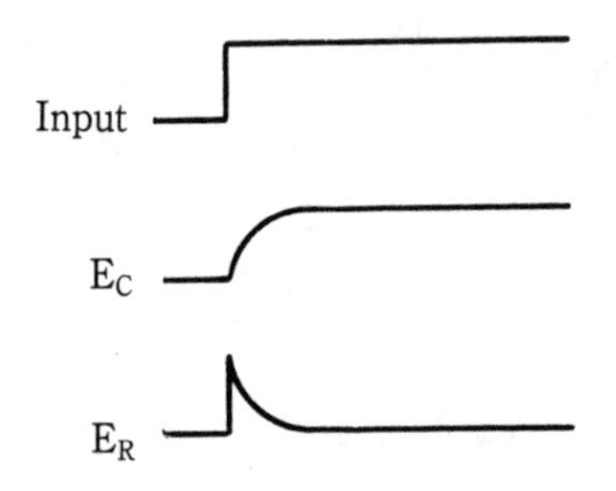

Notice that E_C and E_R at any instant equal the input voltage. Also notice that E_R is the waveform shown in answer a of Q12.

In an inductive circuit like Q13, at the occurrence of a step voltage the inductance will not allow the current to change instantaneously. With no current flow initially, the voltage drop across R is zero therefore the E_L is the applied voltage step. As current builds up, the voltage across R increases, the voltage across L decreases. The voltage will change by 63 percent in one time constant.

$$T = L/R$$

$$\text{in this case } T = \frac{1\text{ mH}}{1\text{ k}\Omega}$$

$$= 1\ \mu\text{ second}$$

Parallel L, R, and C tanks

Inductors and capacitors in parallel exhibit some special characteristics quite useful in electronic circuits. They are used extensively to select frequencies of interest and to reject other frequencies. At low frequencies, capacitive reactance X_C is high because $X_C = 1/2\pi fC$. Inductors at low frequency exhibit a low impedance. $X_L = 2\pi fL$. At resonance $X_C = X_L$. In the parallel circuit of Q14 with the frequency below resonance, then the X_C is greater than the X_L. With smaller X_L the source will find the inductor an easier path for current to flow; hence, answer c is correct.

Question 15 points out some startling concepts of parallel circuits. Examine the circuit of Fig. 5-30.

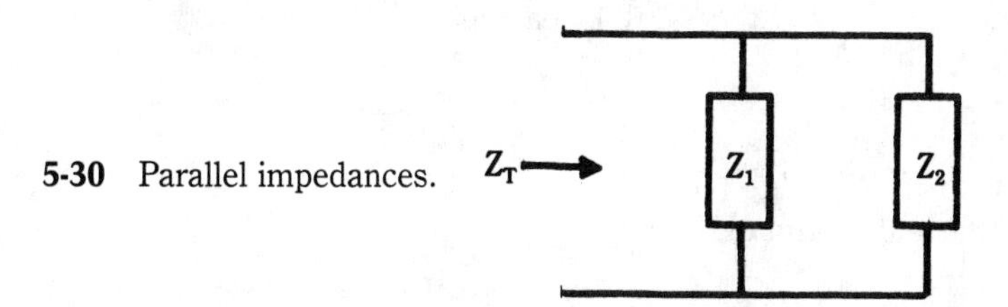

5-30 Parallel impedances.

We know that $Z_T = Z_1Z_2/Z_1 + Z_2$ from work with parallel resistors. But because currents in the impedances are not in phase, neither are the impedances. Therefore, to solve this equation for a real example we have to take into account the phase angles. Using the j operator is one way to do this.

$$Z_T = \frac{(jX_L)(-jX_C)}{jX_L + (-jX_C)}$$

For Q15, $X_L = 30$ and $X_C = 40$

$$Z_T = \frac{(j30)(-j40)}{j30 - j40}$$

$$= \frac{-(j)(j)1200}{-j10}$$

$$\text{but } j = \sqrt{-1} \text{ and } \therefore (j)(j) = -1$$

$$Z_T = \frac{1200}{-j10} \quad (\text{note:} \frac{1}{-j} = j)$$

$$= j120 \text{ ohms}$$

$$= 120 \text{ ohms in phase with the inductor}$$

To illustrate a little further with this example, let's find i_L and i_C and then i_T and finally Z_T, which should be the same as just calculated. A voltage is not given so let's assume 120 volts just for easy division.

$$i_L = \frac{120 \text{ V}}{X_L}$$

$$= \frac{120}{30}$$

$$= 4 \text{ A}$$

$$i_C = \frac{120}{X_C}$$

$$= \frac{120}{40}$$

$$= 3 \text{ A}$$

But these currents are 180° out of phase as illustrated in this phase diagram.

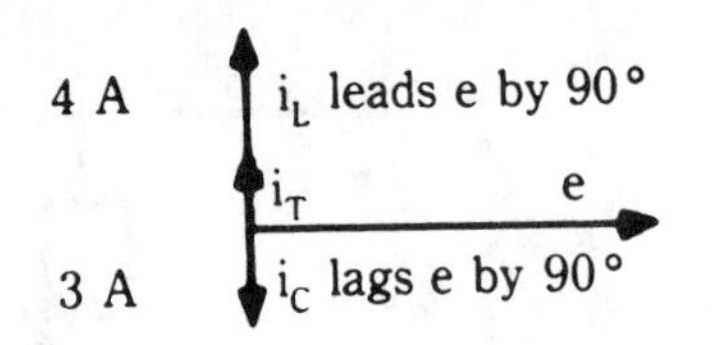

i_T then is the vector sum of 1 ampere.

$$Z_1 = \frac{120 \text{ V}}{1 \text{ A}}$$

$$= 120 \text{ ohms in phase with } i_L$$

Notice that the voltage doesn't alter the result. Try a different voltage and do the same steps if you are unsure of this.

Also notice that if X_L and X_C are equal, i_T would be zero. In a real circuit this is impossible; X_L can equal X_C but some current will have to be drawn from the source because of resistive losses. It does mean, however, that the effective impedance looking into a parallel LC circuit at resonance can be very high. Therefore, a circuit arranged as in Q16 would be a high impedance at or near resonance, but X_L becomes very low at lower frequencies and X_C becomes low at higher frequencies bypassing signals not near resonance. A band of frequencies will be passed near resonance. The circuit is a bandpass filter. Answer d is correct for Q16.

Parallel-series circuit combinations

A series-parallel combination of resistors is shown in Q17. There are an infinite variety of series and parallel combinations possible. This combination is often encountered in electronics and should be familiar. It is also called a bridge circuit. One approach to solving this circuit is to first obtain the voltage drops across individual resistors, then add voltages around a loop from A to B to find the voltage of A with respect to B. The circuit is shown in Fig. 5-31.

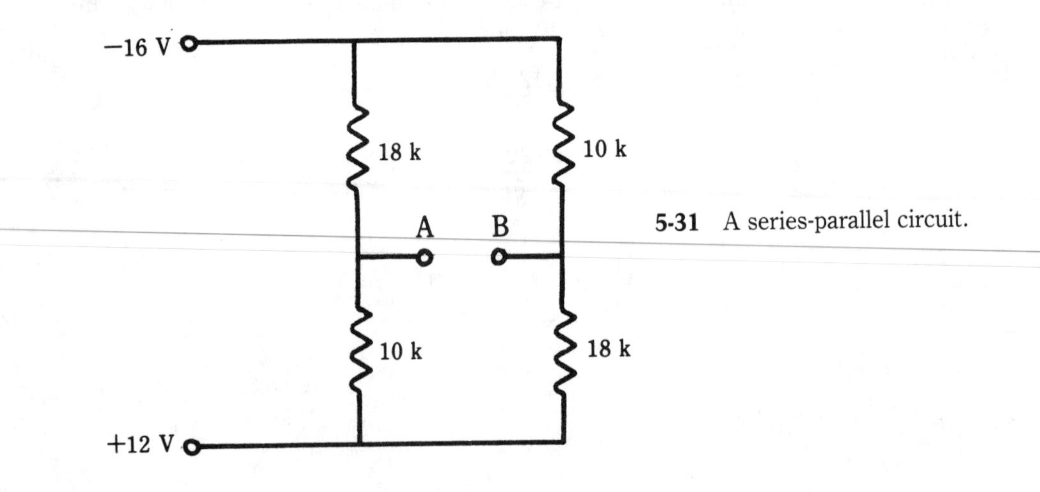

5-31 A series-parallel circuit.

First, let's deal with the +12 volts and −16 volts shown on the schematic. When a supply voltage is shown on a schematic without showing a reference, it is assumed that the voltage is with respect to ground. If we draw in the sources as batteries and connect the grounds, the circuit would look like Fig. 5-32. In this example, the total voltage across each series combination is 16 volts + 12 volts or 28 volts. Each series leg has 28 kilohms total resistance. Hence, the current in each leg is 1 milliampere. Let's multiply each resistance by 1 milliampere and put the voltage drops on each resistor, as in Fig. 5-33. If we sum voltages around the lower loop as indicated by the arrow, the voltage from A to B is:

$$\begin{aligned} V_{AB} &= -10\text{ V} + 18\text{ V} \\ &= +8\text{ V} \end{aligned}$$

Or the voltage at A with respect to ground is $V_{AG} = -10\text{ V} + 12\text{ V} = 2\text{ V}$ and B is $V_{BG} = -18\text{ V} + 12\text{ V} = -6\text{ V}$ and hence, because A is +2 V above ground and B is 6 V below ground, A is 8 volts above B.

Another possible parallel series combination of resistors is shown in Q18. In this circuit the actual value of R1 can be found by dividing 50 volts by (10 mA + 20

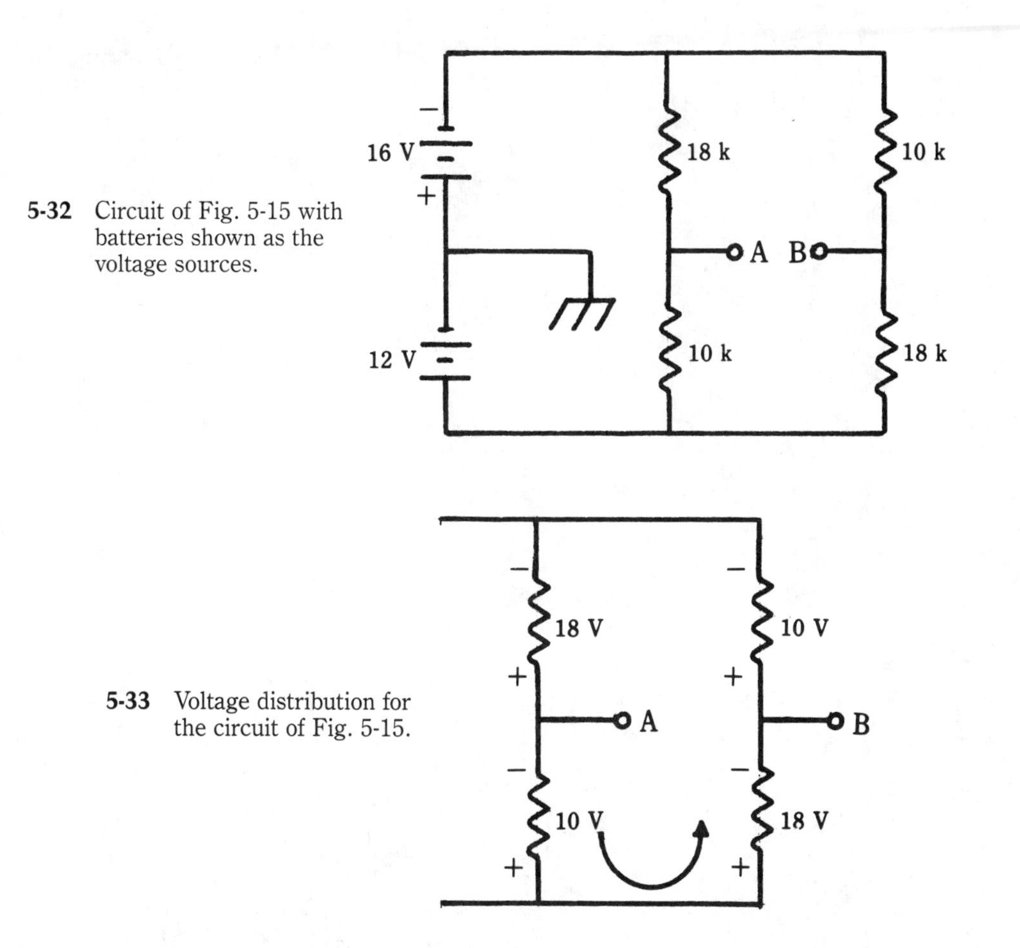

5-32 Circuit of Fig. 5-15 with batteries shown as the voltage sources.

5-33 Voltage distribution for the circuit of Fig. 5-15.

mA) the total current flowing through R1. This value is 1670 ohms (1.67 k). A 20 percent, 1500-ohm resistor can vary from 1500 – 0.2(1500) to 1500 + 0.2(1500) or 1200 to 1800 ohms. Because 1.67 k is within this range answer b is correct.

Additional reading

Larch, E. Norman: *Fundamentals of Electronics*, John Wiley.

Lease, Alfred: *Basic Electronics*, The Bruce Publishing Company, 1965.

Mandl, Matthew: *Fundamentals of Electronics*, Prentice-Hall, 1960.

Edwards, John: *Exploring Electricity and Electronics with Projects*, TAB Books, 1983.

6
Semiconductors

THIS CHAPTER CONTAINS QUESTIONS ON DIODES, TRANSISTORS, AND other semiconductor devices. This review covers operation, safety precautions, and testing of these devices.

Quiz

Q1. A forward-biased diode made of germanium will have a voltage drop such that the anode will be:

a. positive with respect to the cathode.
b. negative with respect to the cathode.

Q2. The symbol for a zener diode is:

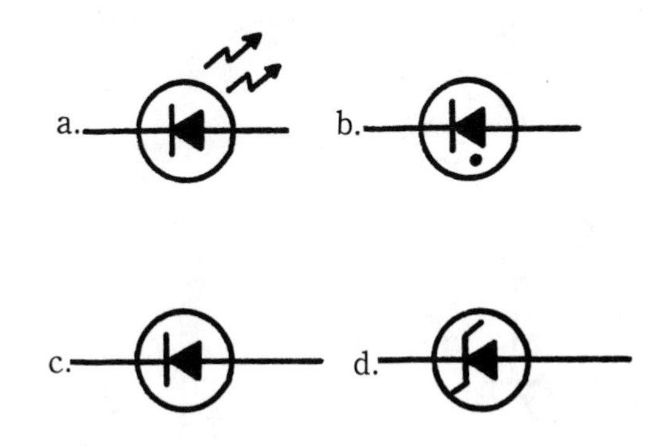

Q3. In the circuit of Fig. 6-1 the voltage at A with respect to B should be:

a. positive.
b. negative.
c. alternating.

Q4. If the zener diode opens in the circuit of Q3, the output voltage will:

a. increase.
b. decrease.
c. remain the same.

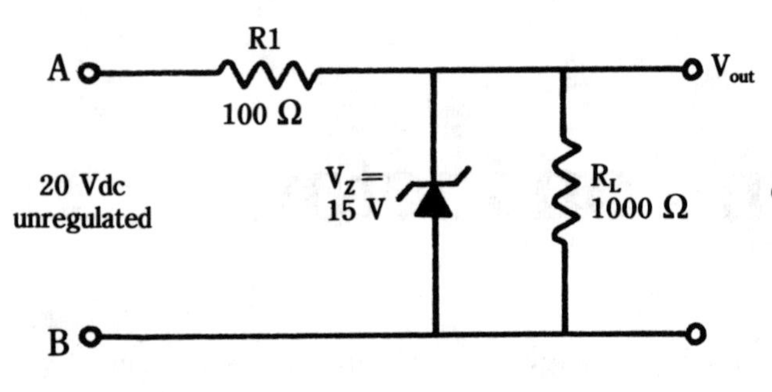

6-1 Zener regulator for Q3.

Q5. The bipolar transistor in Fig. 6-2 is:

a. an NPN, biased correctly.
b. an NPN, biased incorrectly.
c. a PNP, biased correctly.
d. a PNP, biased incorrectly.

Q6. A transistor is biased at saturation with a sine wave signal input to the circuit. The output clips 180° of the input. This is:

a. class A operation.
b. class B operation.
c. class C operation.
d. None of these.

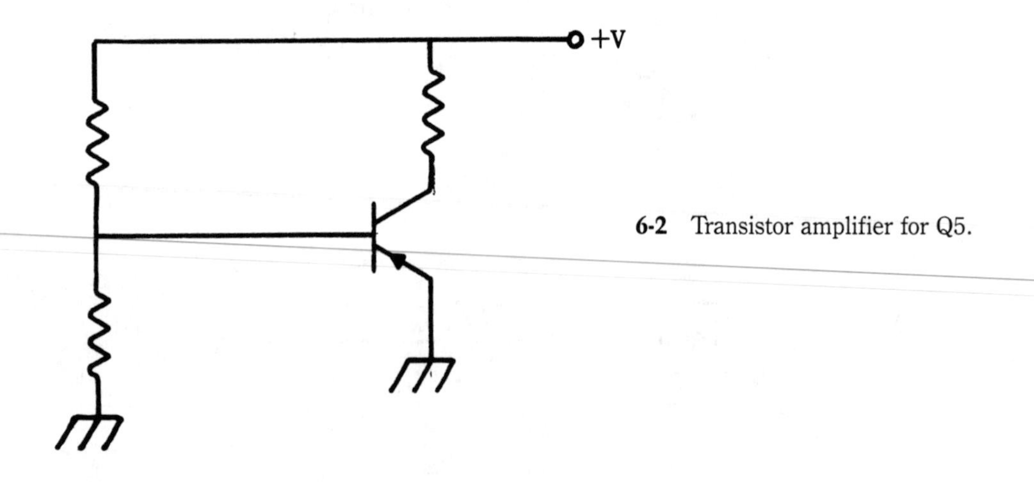

6-2 Transistor amplifier for Q5.

Q7. Figure 6-3 is a/an:

a. common-base amplifier.
b. common-emitter amplifier.
c. common-collector amplifier.
d. None of these.

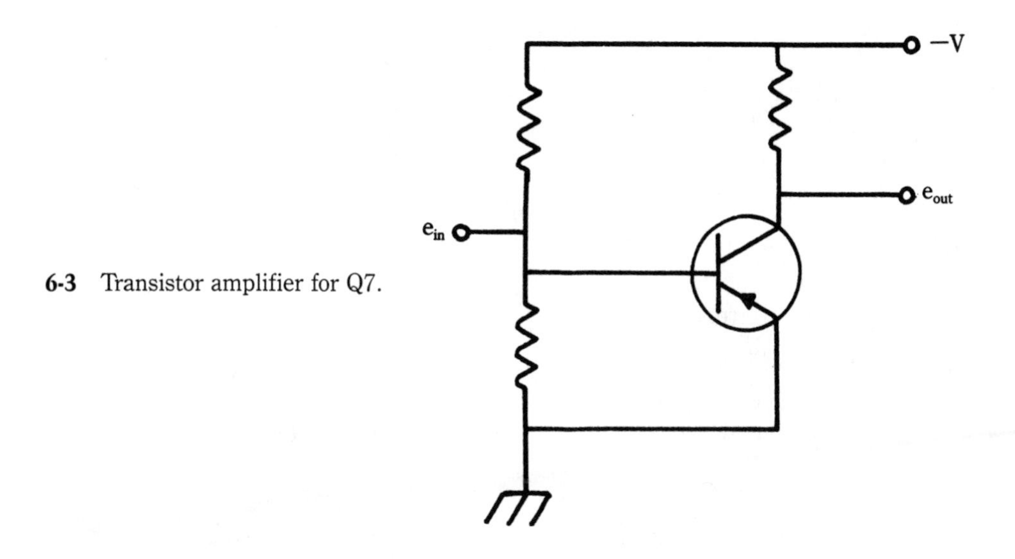

6-3 Transistor amplifier for Q7.

Q8. Suppose in the circuit of Fig. 6-4 that the generator creates a voltage signal that produces a signal current i_e of 2 mA. The collector signal current i_c will be:

a. approximately 2 mA.
b. beta times i_e.
c. since beta = 1 in common base circuits a and b are both correct.
d. None of these.

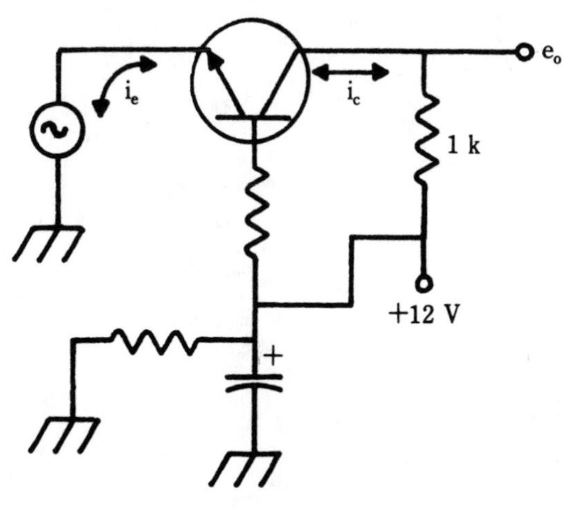

6-4 Currents in common-base amplifier, Q8.

Q9. If in the circuit of Fig. 6-4 it takes 0.01 volt of signal to create i_e of 1 mA the voltage gain is:

a. undeterminable.
b. unity.
c. 100.
d. less than one.

Q10. An ohmmeter on the ×100 scale is being used to measure a transistor. The red lead is positive with respect to the black lead. A low reading is measured when the red lead is on the base and the black lead is on the collector of a PNP transistor. This measurement is:

a. normal.
b. not normal.

Q11. A unijunction transistor has:

a. an emitter.
b. a negative resistance region.
c. three leads.
d. All of the above.

Q12. The symbol in Fig. 6-5 is:

a. an N-channel unijunction.
b. an N-type SCR.
c. an N-type SCN.
d. an N-type MOSFET.

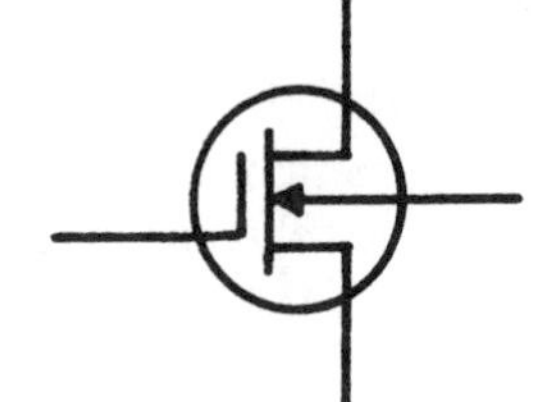

6-5 Solid-state device for Q12.

Q13. In the symbol of the SCR in Fig. 6-6 terminal C is:

a. the cathode.
b. the anode.
c. the gate.
d. None of these.

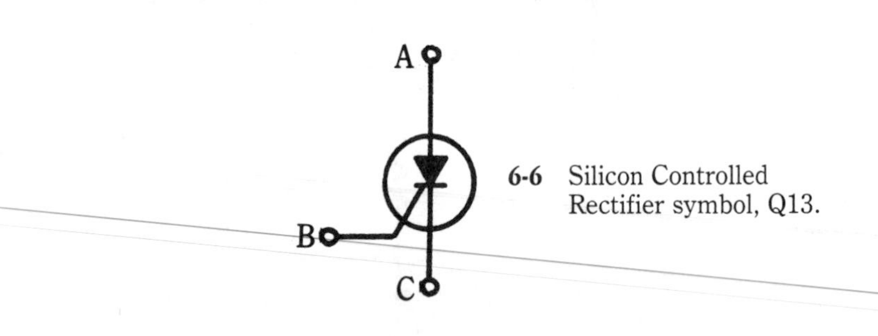

6-6 Silicon Controlled Rectifier symbol, Q13.

Q14. Figure 6-7 is the symbol of:

a. a triac.
b. a silicon-controlled switch.
c. a diac.
d. None of these.

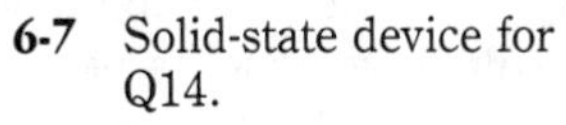

6-7 Solid-state device for Q14.

Diodes

The following circuit (Fig. 6-8) is biased such that current will flow easily through the circuit. If the positive terminal of the battery is connected through circuitry to the anode of a diode and the negative terminal is connected through circuitry to the cathode, the diode is forward biased. Electrons flow as indicated by the arrow and hence voltage drops are as shown. This is opposite the arrow portion of the diode symbol. The answer to Q1 is therefore a. If the diode is turned around (as shown in Fig. 6-9), the voltage drops are still in the same direction. However, the diode is now reverse biased. Therefore, current is quite small. The voltage drop on the resistor is almost zero, hence the back-bias on the diode is nearly equal to the battery voltage, 10 volts. When the diode is forward biased the magnitude of its voltage drop is approximately 0.6 volt (silicon diode), or 0.3 V (germanium).

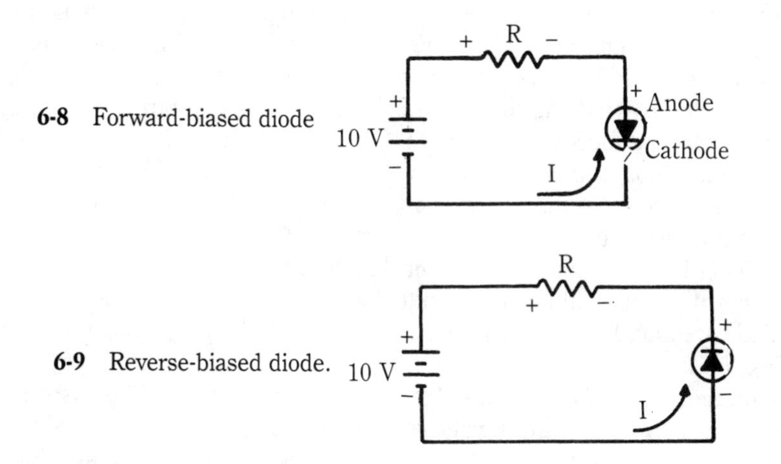

6-8 Forward-biased diode

6-9 Reverse-biased diode.

When replacing diodes, four things are important to know. First, I_{Fmax}, the forward current maximum rating is important. Suppose R in Fig. 6-8 is 10 ohms ±10 percent. I_{max} will occur when R is minimum and the diode is forward biased.

$$I_{max} = \frac{10\text{ V} - 0.6\text{ V (voltage drop across the diode)}}{10 - 10(0.10) \quad \text{(minimum resistance)}}$$

$$= 1.044\text{ A}$$

Select a diode in this case with I_{Fmax} greater than 1.044 amps. Second, PIV (the peak inverse voltage rating) is important to know if the diode is to be reverse biased. Suppose the circuit of Fig. 6-10 is to be built. What should PIV be?

The highest voltage in the reverse direction occurs when the capacitor is charged to peak value and the input voltage goes to its peak value with terminal A negative with respect to B. We might redraw the circuit at that instant to give Fig. 6-11.

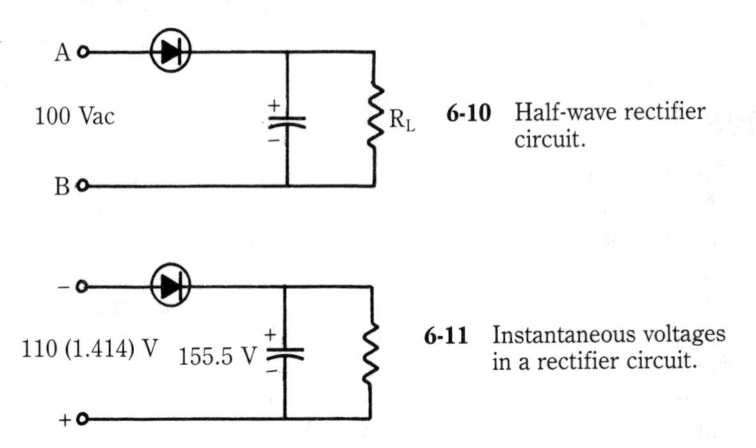

6-10 Half-wave rectifier circuit.

6-11 Instantaneous voltages in a rectifier circuit.

Hence, the reverse voltage applied to the diode is 110 (1.414) + 110 (1.414) = 311 volts. In many places the 110-volt ac line voltage is closer to 120 volts ac. If the above source voltage was from such a power system then the actual applied inverse voltage could be as high as 340 volts. In that case, the selected diode must have a PIV (peak inverse voltage) greater than 340 volts. The third important factor is what the diode material is. Is it germanium, silicon, or some other material? The fourth important factor is the physical characteristics. Will it fit? If it's a stud-mounted diode, does the stud need to be insulated? Is the stud the anode or the cathode? All other physical factors need to be considered. An exact replacement takes care of all these considerations (provided that the exact replacement really is just that, and isn't mismarked or a "bad" part).

Another diode (useful for its special characteristics) is the zener diode. In the zener diode symbol the cathode portion is shown as a bent line resembling the letter Z (Fig. 6-12).

I recommend, ETA/s, the symbol on the right in Fig. 6-12. The answer to Q2 is d. Zener diodes are important in voltage regulations. These diodes have a range of operation when reverse biased, that exhibits almost constant voltage for a wide range of currents. To reverse bias a zener a positive voltage must be sent to the cathode and a negative sent to the anode. In Q3 this means A must be positive with respect to B, hence answer a is correct.

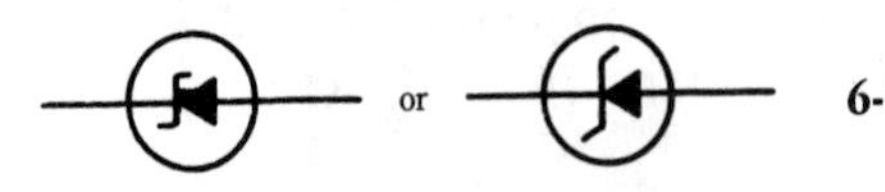

6-12 Zener diode symbols.

In the circuit of Fig. 6-13 (with a zener diode), normal operation is such that the voltage drop across R1 is $V_B - V_Z$. The current through $R_1(I_T)$ is $I_Z + I_L$ where I_Z is the zener current (electron current shown on drawings) and I_L is the load current. As long as the operation is in the normal range I_T is constant and changing the load current I_L changes I_Z inversely. Let's put in some numbers and explore this further. Let $V_B = 10$ volts, $V_Z = 5$ volts and R1 = 1000 ohms. I_T must then be constant at 10 V − 5 V/1000 = , which is equal to 5 milliamps. Suppose $R_L = 2000$ ohms. Then $I_L = 5/2000 = 2.5$ mA. If $I_T = 5$ milliamps and $I_L = 2.5$ milliamps then I_Z must equal 2.5 mA. Suppose that R_L opens up. Then I_L is surely 0 mA. But I_T must still be 5 mA. Let's change R_L now to 1000 ohms. Then I_L wants to be 5 mA and I_Z wants to be 0 milliamps. Or, the zener diode is no longer in the zener region. It is starved for current. A further increase in the load takes the circuit out of normal operation. For example, suppose R_L changes to 500 ohms. Now the circuit current I_T must increase because $I_1 + R_L = 1500$ ohms and 10 volts/1500 ohms = 6.67 mA. The load voltage is now 6.67 mA × 500 = 3.33 volts. The zener is now off or not "broken down."

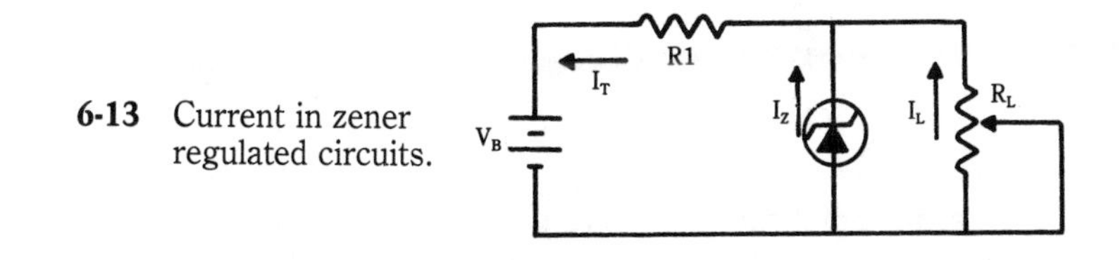

6-13 Current in zener regulated circuits.

In Q4 a circuit is present and a trouble introduced. The zener is opened. In that case the circuit now looks like Fig. 6-14.

$$V_{OUT} = 20\ \text{V}\frac{1000}{1000 + 100} = 18.2\ \text{volts}$$

Therefore, V_{OUT} increased from 15 volts to 18.2 volts.

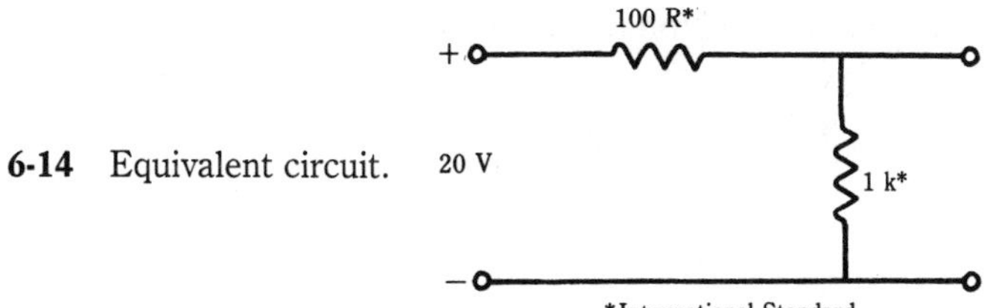

6-14 Equivalent circuit.

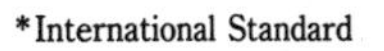

The subject of diodes usually covers tunnel diodes. The symbol is shown in Fig. 6-15. This diode, when slightly forward biased, exhibits negative resistance. That is when the voltage is increased the current decreases. It is possible to build an oscillator using this characteristic with the circuit shown in Fig. 6-16.

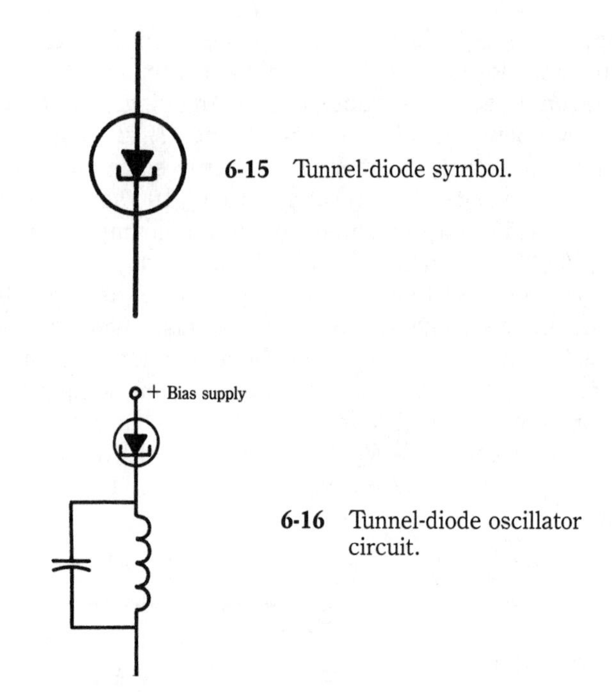

6-15 Tunnel-diode symbol.

6-16 Tunnel-diode oscillator circuit.

However, because of difficulties in keeping the diode in the negative resistance range, and other problems, the tunnel diode has not been used much in practical circuits. On the other hand, another diode, the *varicap* diode or varactor is very useful. This diode is normally reverse biased. Because a reverse biased junction is essentially electrons separated by a dielectric, it has capacitance. Because the capacitance varies (separation of the carriers) as the reverse voltage varies, the diode is useful in construction of voltage-variable oscillators. Special diodes are also available for fast switching, noise generation, and other special needs.

Three-lead semiconductors

A very useful three-lead semiconductor device is the bipolar transistor. There are many varieties and types of bipolar transistors. The symbols for an NPN and for a PNP are shown in Fig. 6-17.

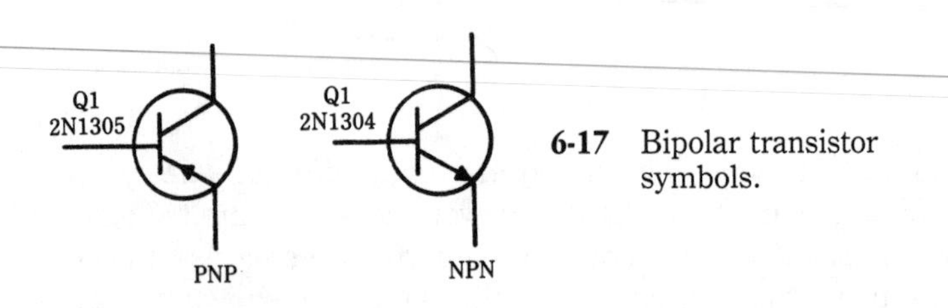

6-17 Bipolar transistor symbols.

Transistor amplifiers

Bipolar transistors are used as amplifiers. To make them operate they must be properly biased. To amplify, the collector-base junction must be reverse biased and the base-emitter forward biased. A forward biased junction is very much like a forward biased diode. In the circuit of Fig. 6-18 the base is tied through the resistor to the negative terminal of the battery, while the emitter is tied to the positive terminal of the battery. The arrow is in the direction of conventional current flow; that is, opposite to electron flow. If the battery is connected to cause electrons to flow in a direction opposite to the arrow, the junction is forward biased. A forward biased silicon junction has about 0.6-volt drop. A germanium junction has about 0.3-volt drop if biased in the forward direction.

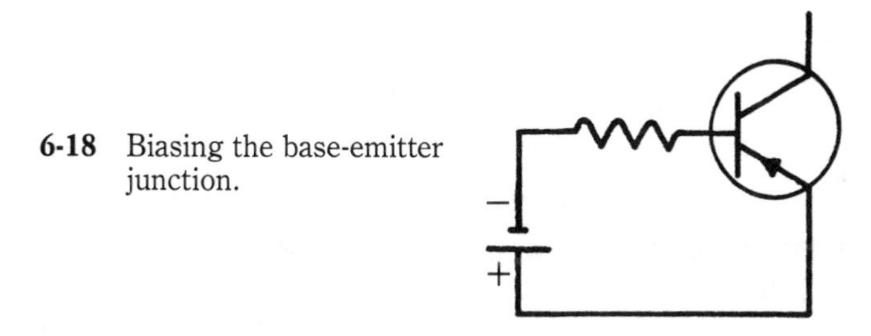

6-18 Biasing the base-emitter junction.

The collector-base junction of Fig. 6-19 is reverse biased. (Think of the collector lead also having an arrow pointing in, like the emitter.) Because the negative battery terminal is connected to the collector through the resistor and the positive battery terminal is connected to the base, the electron flow would be in the direction of the arrow; if there was one, and if current flowed. But this is reverse bias so no current flows. Putting the two concepts together in Fig. 6-20 gives a properly biased amplifier.

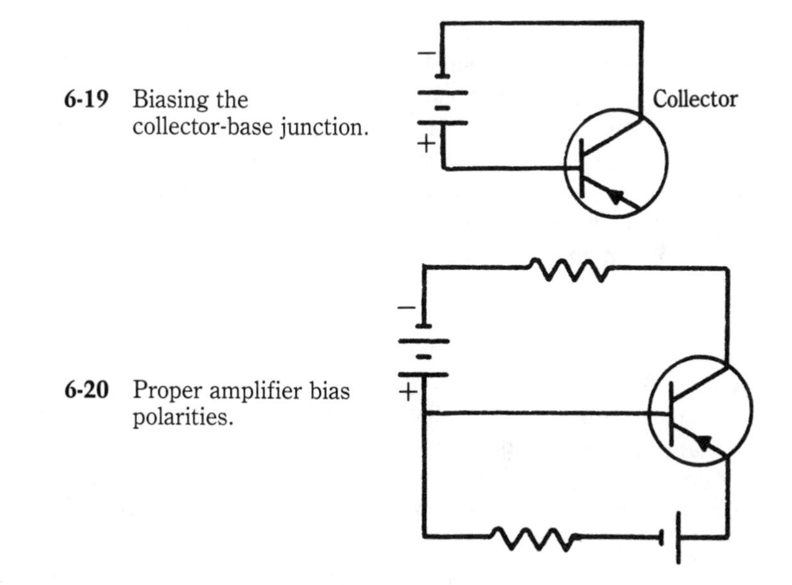

6-19 Biasing the collector-base junction.

6-20 Proper amplifier bias polarities.

With the base-emitter conducting, electrons will be "injected" into the base. These electrons will cause current to flow in the collector lead also. This current will be h_{FE} times the base current. The h_{FE} is the dc current gain. In equation form:

$$I_C = h_{FE} I_B$$

Let's look at the circuit of Q5 redrawn as in Fig. 6-21 with voltage drops shown.

Notice that the voltage drops across the power supply voltage divider are as shown. These drops are such that they would *reverse* bias the base-emitter junction and *forward* bias the collector-base junction. This is incorrect for proper operation of the amplifier. Hence, this is an improperly biased PNP transistor.

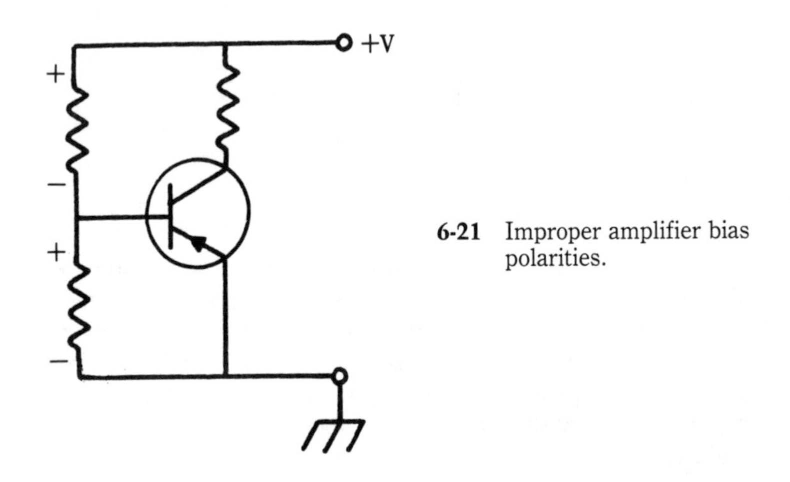

6-21 Improper amplifier bias polarities.

With the circuit of Fig. 6-22, the base-emitter bias can be adjusted from highly positive to highly negative. If the bias is highly negative the base-emitter junction is reverse biased. Decreasing the negative voltage to the point where signal starts to affect the output on the positive signal peaks would be approaching Class C operation. That is, output signal change for less than 180° of input signal. At some point in decreasing the negative voltage, the output signal would vary for 180° of the input signal. That bias is called Class B bias. Increasing the bias positively would eventually reach an area of bias where signal output would change for the entire 360° of input signal. This area is Class A operation. Further increasing bias voltage in the positive direction would establish another point where for half the input cycle the output would not vary because the transistor would be saturated for that half cycle. This is another Class B operational bias. Increasing bias more positively would reduce output signal variation to less than 180° of the input signal. This would again be Class C operation. Eventually, if the bias is positive enough the signal could not bring the transistor out of saturation. The answer to Q6 is b. Digital circuits are normally biased either into saturation or at cutoff, that is, on or off.

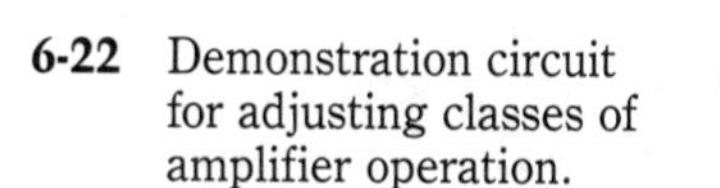

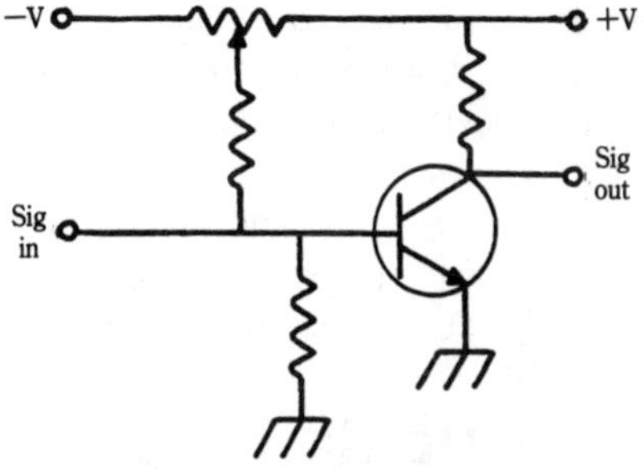

6-22 Demonstration circuit for adjusting classes of amplifier operation.

CE, CC, or CB amplifiers

Amplifiers are sometimes classified by the terminology common-emitter, common-collector or common-base amplifiers, depending on the common connection between the input signal and the output signal. For example, if the base is common to the input signal and to the output signal, the amplifier is a common-base amplifier. Because it is almost an unwritten rule that signal flow is depicted on schematics from left to right, a common-base amplifier is often depicted as in Fig. 6-23.

No matter how the circuit is drawn, if the input and output signals show the base as their common terminal, it is a common-base amplifier. These amplifiers are useful when a high-input impedance is needed.

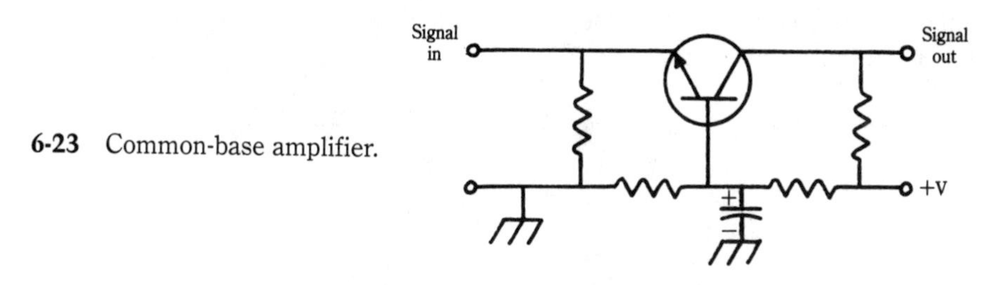

6-23 Common-base amplifier.

Common-collector amplifiers are often called emitter followers because the output signal taken from the emitter follows almost one for one what the base signal is. Figure 6-24 shows an emitter-follower schematic or common-collector amplifier schematic. The signal is developed from base to ground and from emitter

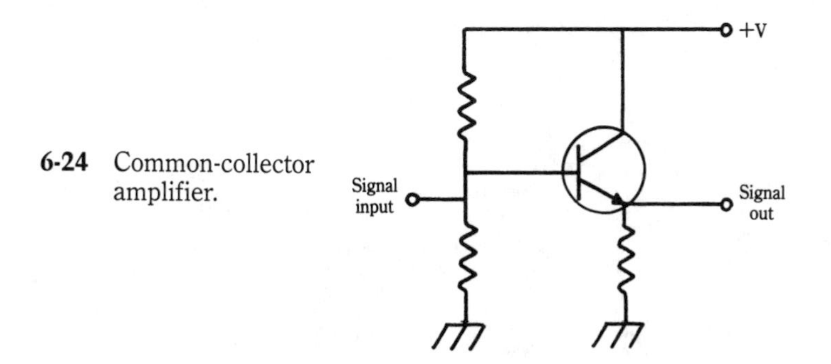

6-24 Common-collector amplifier.

to ground. But *signal-wise* the collector is virtual ground. Collector voltage does not vary with either input or output signal. Therefore, the collector is "common" to the input and to the output signals.

The common-emitter amplifier is the most commonly used transistor amplifier because it has current gain and voltage gain. A common-emitter amplifier is shown in Fig. 6-3. In this case, ground happens to be also common for the emitter as well as the input signal and the output signal.

Circuit calculations

Because common-emitter amplifiers are widely used, they are discussed first. If you see the circuit of Fig. 6-25 in the certification test and you are asked what the collector dc voltage is, could you determine the value? If you can, you should not have trouble with bias problems on the exam. If, in addition to the information given in the circuit, you are also told this is a Class A amplifier, you should be able to get close to the answer without any calculation because to amplify a signal linearly for 360° of the input signal, the transistor should be biased somewhere in the middle, between saturation and cutoff. In this case, if the transistor was cut off the collector voltage would be 10 volts. If it were saturated the collector voltage would be nearly 0 volts. For an amplifier you should expect the collector to be about halfway between these extremes, at 5 volts. But with the information contained in the drawing the collector voltage can be calculated. Start by calculating the base to ground voltage. Because the base draws very little current, this voltage will be determined by the voltage divider R1, R2.

$$V_{BG} = \text{voltage base to ground} = \frac{1.2\text{ k}\Omega}{1.2\text{ k}\Omega + 8.2\ \Omega} \times (10\text{ V})$$

$$= \frac{12\text{ V}}{9.4}$$

$$= 1.28\text{ V}$$

If the transistor is silicon, the base-emitter voltage drop will be about 0.7 volt. Hence the voltage from emitter to ground should be:

$$V_{EG} = 1.28 - 0.7$$
$$= 0.58\text{ V}$$

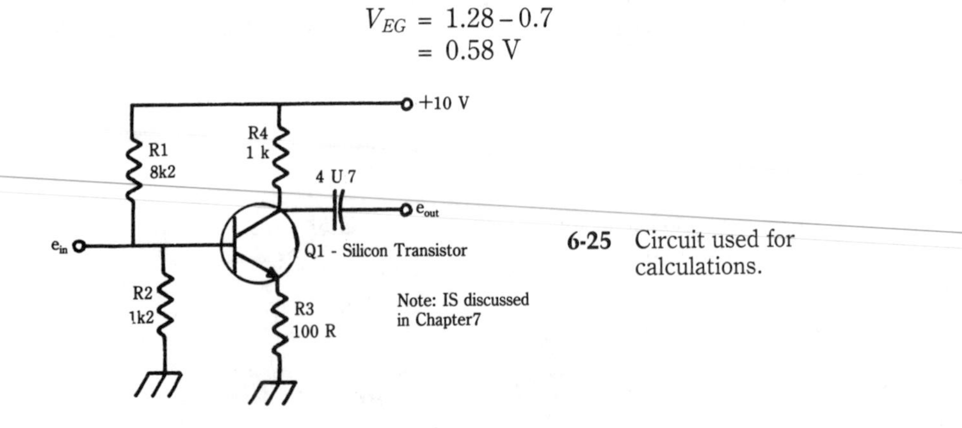

6-25 Circuit used for calculations.

The emitter current must equal:

$$I_E = \frac{0.58\text{ V}}{100\ \Omega}$$
$$= 5.8\text{ mA}$$

Because the emitter current and collector current are almost identical then:

$$I_C = I_E = 5.8\text{ mA}$$

This is a good approximation. The collector voltage to ground then is:

$$V_{CG} = 10\text{ V} - I_C\text{R4}$$
$$= 10\text{ V} - (5.80\text{ mA})\ 1\text{ k}\Omega$$
$$= 4.2\text{ volts}$$

Suppose a signal is applied at the base with respect to ground, could you calculate the output signal? Say that e_{in} is 20 mV P-P; that is, the base voltage varies from 1.29 up to 1.3 volts and down to 1.27 volts. With no bypass capacitor on the emitter, its voltage will also vary by 20 mV P-P (or near enough, anyway). Therefore, I_E will vary by:

$$\Delta I_E = i_e = \frac{20\text{ mV}}{100\ \Omega} = 0.2\text{ mA}$$

$$\text{and } \Delta I_C = \text{i}_\text{c} = i_e$$
$$\text{hence: } e_O = i_c \times 1\text{ k}\Omega$$
$$= (0.2\text{ mA})\ (1\text{k}\Omega)$$
$$= 0.2\text{ volts P-P}$$

The voltage gain:

$$A_v = \frac{e_o}{e_{in}}$$
$$= \frac{0.2}{20\text{ mV}}$$
$$= 10$$

In this common-emitter circuit the voltage gain is easier to figure by noting that because the signal current in the emitter resistor is the same as the collector signal current the voltage gain must be the same as the ratio of collector resistance to emitter resistance:

$$A_v = 1\text{ k}/100 = 10$$

With the emitter bypassed, however, none of the above holds true. Because the emitter voltage stays constant, there is no "built-in" degeneration. In other words, all of the signal results in a base current swing. The collector current change is then the ac current gain, h_{fe}, times the base current change. These are both dependent on the transistor parameters. Voltage gains of 100 are common and much higher is possible. In this configuration both current gain and voltage gain are exhibited.

The common-base circuit in Fig. 6-26 is the same as that in Q8. The currents drawn are dc currents; therefore, they are capital letters. Also they are drawn the way electrons would flow in the circuits. Many references draw all the currents either into or out of the transistor and let the signs take care of direction. That is, if a current is drawn into the collector and it is later found that the current has a negative sign, it means the current flows out. Anyway, the way they are drawn here is the way electrons would flow in the properly biased transistor. Because electrons can't disappear, the current in must equal the current out.

$$I_E = I_B + I_C$$

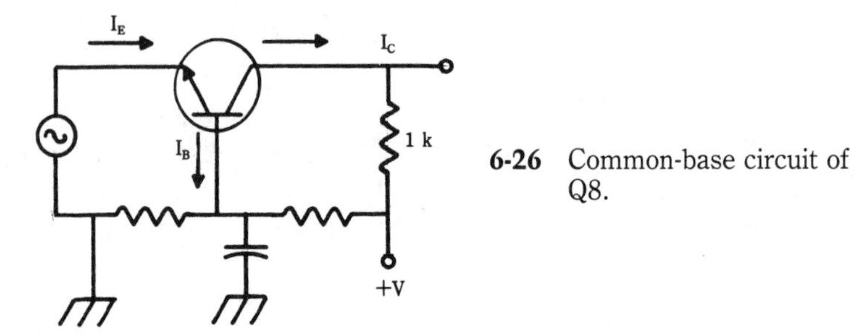

6-26 Common-base circuit of Q8.

Or rearrange the equation:

$$I_C = I_E - I_B$$

Notice that because I_B is very small, I_C is always slightly less than I_E. If the signal input voltage causes a slight increase in I_E, I_B, and I_C will increase. If I_E decreases slightly I_B and I_C will also decrease. In other words, signal currents will behave much the same as dc currents. To indicate signal voltages and currents, use lowercase letters as in Fig. 6-27.

$$i_e = i_c + i_b$$

or rearranging $i_c = i_e - i_b$ with i_b very very small $i_c \approx i_e$; hence, if the signal voltage causes a signal current i_e of 2 milliamperes the collector current is almost 2 milliamperes. Answer a is correct for Q8. If a signal of 0.01 volt creates a current of 1 milliampere, most of the 1 milliampere flows to the collector and hence through the 1 k collector resistor resulting in a signal voltage change of 1 volt. The voltage gain then is:

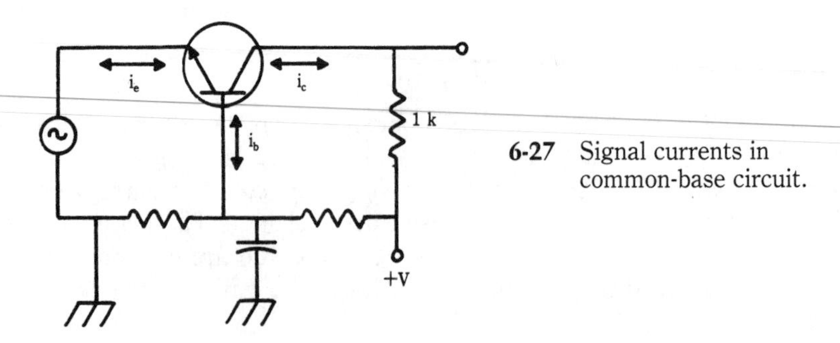

6-27 Signal currents in common-base circuit.

$$A_v = \frac{1\text{ V}}{0.01\text{ V}}$$
$$= 100 \text{ (Answer to Q9 is c.)}$$

A common-collector circuit is shown in Fig. 6-28. Again, the base to ground voltage can be calculated ignoring the base current.

$$V_{BG} = \frac{1\text{ k}\Omega}{2\text{ k}\Omega}(-10)$$
$$= -5\text{ V}$$

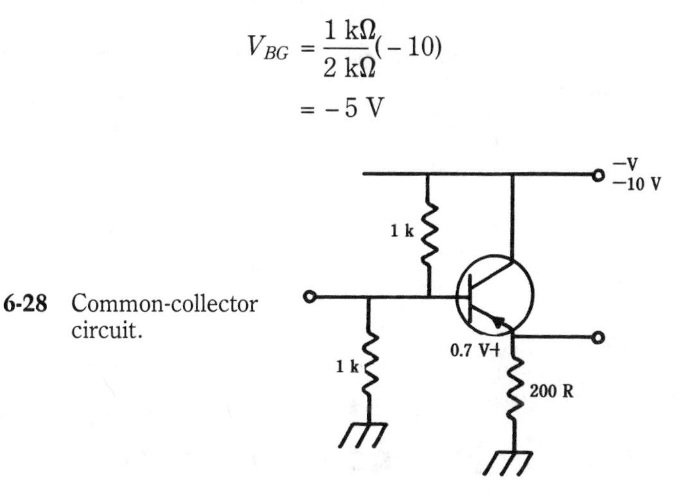

6-28 Common-collector circuit.

If the transistor is made of silicon the base-emitter drop is 0.7 volt, hence:

$$V_{EG} = -5\text{ V} + 0.7$$
$$= -4.3\text{ V}$$

and from this

$$I_E = \frac{V_{EG}}{\text{R}} = \frac{-4.3}{200} = 21.5\text{ mA}$$

Note that the input signal causes the base voltage to vary around the 5-volt level. For example, suppose the input voltage causes the base voltage to change from -5 volts to -6 volts. If that's the case, the output voltage goes from -4.3 to -5.3; still 0.7 volt away from the base voltage. If the input causes the base voltage to go to -4 volts the output changes to -3.3 (still 0.7 volt more positive than the base). Hence, while the input signal is two volts peak-to-peak the output is two volts peak-to-peak. This example ignores some things like the nonlinearity of the transistor. However, the output signal voltage will be only slightly less than the input signal voltage.

A_V is almost 1

The base current change required to change the emitter current is:

$$\Delta I_B = \Delta I_E / h_{fe}$$

The change in emitter current:

$$\Delta I_E = \frac{5.3}{200} - \frac{3.3}{200} = \frac{2}{200} = 10\text{ mA}$$

If the transistor's small signal current gain is 50 (not an exceedingly high current gain), then:

$$\Delta I_B = \Delta I_E / h_{fe}$$
$$= \frac{10 \text{ mA}}{50}$$
$$= 0.2 \text{ mA}$$

You can see from these examples that the following conditions exist in the various amplifier configurations.

Circuit configuration	A_V Voltage gain	A_I Current gain
Common emitter	>1	>1
Common base	>1	≈1
Common collector (emitter-follower)	≈	>1

Measuring diodes and transistors

An ohmmeter is often used to check diodes and transistors. A semiconductor junction when forward biased is a low resistance and when reverse biased it is a high resistance. Not all meters have the positive terminal of the internal battery (or voltage source) attached to the red lead. If the red lead is positive when it is connected to the anode of a semiconductor diode and if the black negative lead of the ohmmeter is attached to the cathode lead, the diode is forward biased provided that the voltage is large enough (Fig. 6-29). If the junction is good the resistance will be low, maybe 10 ohms. If the leads are reversed the resistance will be high, maybe 10 kilohms.

In measuring transistors the same sort of thing occurs. If the junctions are looked at as if they were diodes, the equivalents of Fig. 6-30 will prevail. If the positive red lead is on the base of the PNP for example, and the black lead on the collector, the junction is reverse biased and the reading should be high. Question 10 stated the reading was low, hence b "not normal" is the correct answer.

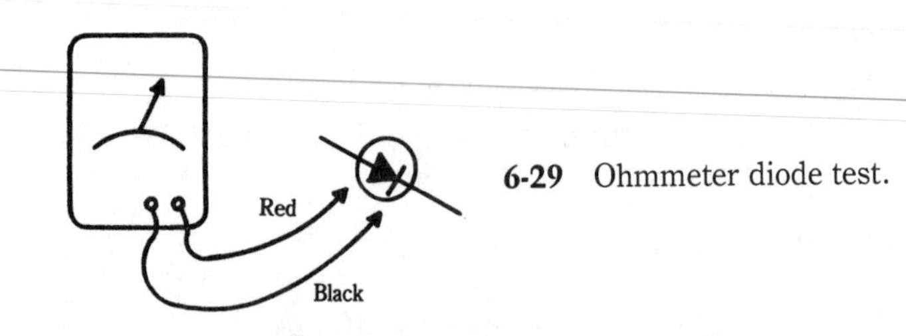

6-29 Ohmmeter diode test.

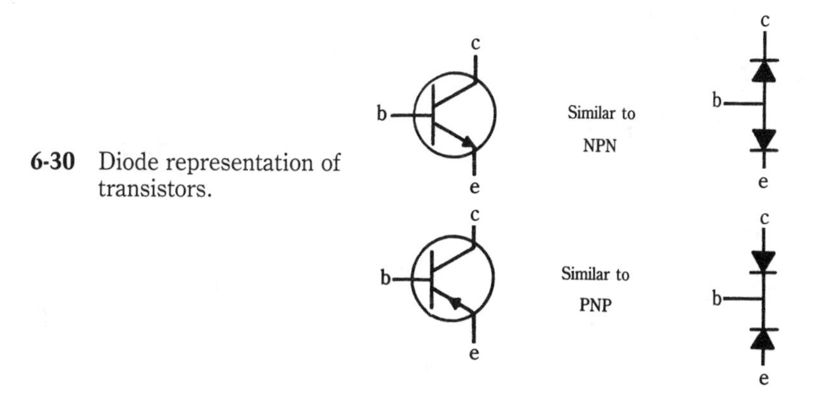

6-30 Diode representation of transistors.

Other semiconductor devices

There are many types of semiconductor devices. This section covers some of these devices. The symbol for *unijunction transistor* is shown in Fig. 6-31. The N material between base 1 and base 2 acts like a conductor with some resistance. If current is flowing in the base 1 emitter circuit, however, the resistance is very low between base 1 and base 2. The base 1 emitter has a region (when forward biased) of negative resistance. Like the tunnel diode, in this region increases of forward bias result in decreased current. This characteristic is useful in building a simple oscillator. Because the unijunction transistor has an emitter, and it has this area of negative resistance, and it has three leads, the answer to Q11 is d.

6-31 Unijunction transistor symbol.

Another very useful semiconductor device is the *junction field-effect transistor* (JFET). There are two kinds of JFETS: the N-channel and the P-channel. Each has a gate, a drain, and a source (Fig. 6-32). While JFETs are useful they have not been applied as extensively as their cousins, the *isolated gate field-effect transistors* (IGFET). This transistor is also known as a MOSFET (*metal-oxide semiconductor FET*). The JFET (like the unijunction) consists of a P-doped or N-doped silicon bar that acts like a resistance between the source and drain. The gate is diffused into the bar making a PN junction between the gate and the bar. If the junction is forward biased and it increases current in the bar between the source and drain, the transistor is operating in the enhancement mode. If reverse bias decreases current, the transistor current is being depleted. The transistor is said to be operating in the depletion mode. The symbols in Fig. 6-33 reveal that the correct answer to Q12 is d.

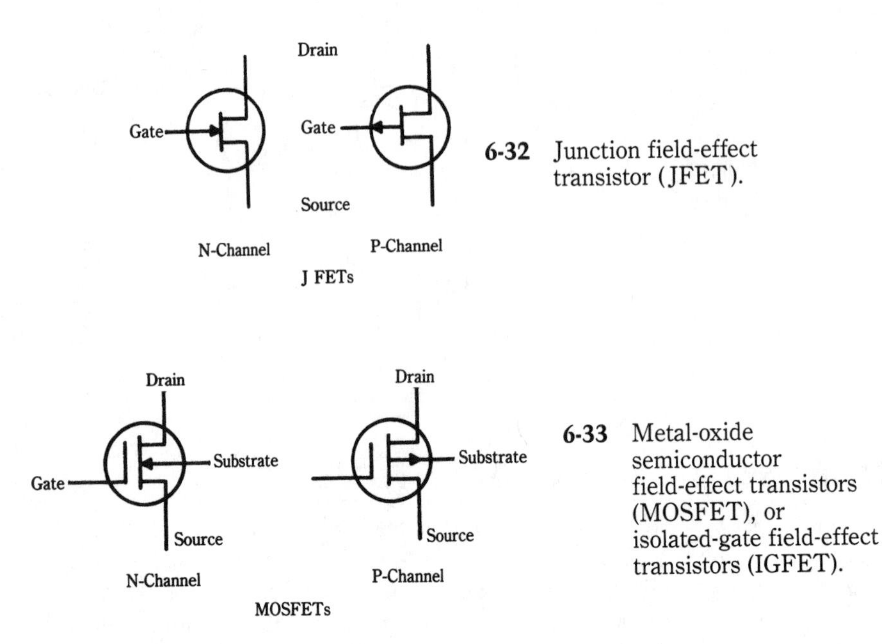

6-32 Junction field-effect transistor (JFET).

6-33 Metal-oxide semiconductor field-effect transistors (MOSFET), or isolated-gate field-effect transistors (IGFET).

The MOSFET is useful because it uses small amounts of power and it has a very high input impedance. By combining N-channel and P-channel FET devices on a single substrate, a pair of complementary FETs are created. A whole family of devices using these techniques have been devised and gained widespread use in computer technology. These are called CMOS devices (pronounced see-moss), see Fig. 6-34.

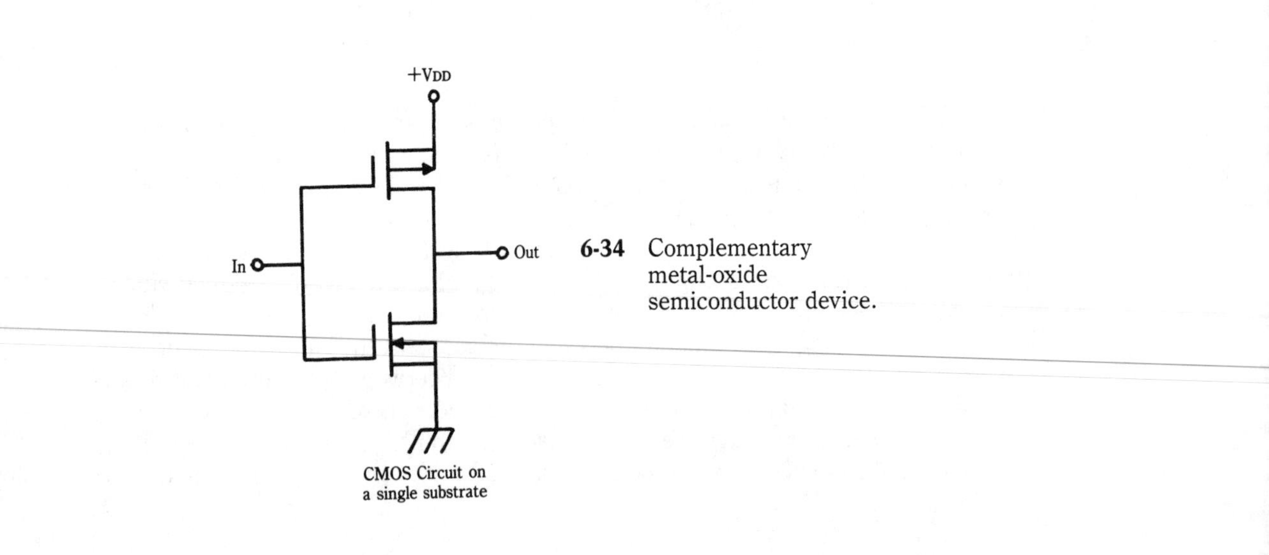

6-34 Complementary metal-oxide semiconductor device.

Thyristors

A large group of semiconductor devices having three or more junctions have been developed. These are called *thyristors*. Some thyristor symbols are shown in Fig. 6-35. The SCR behaves like a gas-filled triode, except current in the gate starts conduction instead of voltage on a grid. With positive voltage at the anode (with respect to the cathode), the SCR will not conduct normally. Application of positive voltage on the gate (with respect to the cathode) causes gate current to flow. The gate current "turns on" the SCR. Anode current flows. If the anode-to-cathode voltage is above a certain breakover voltage, the SCR can conduct without gate current. This would not be allowed in most applications, because the advantage of the SCR is in controlling the turn-on with a small gate current. The SCR symbol shows that the correct answer to Q13 is a (the cathode). Notice that positive voltage applied to the cathode of the SCR (with respect to the anode) will not result in current flow. The SCR acts like a back-biased diode regardless of gate current.

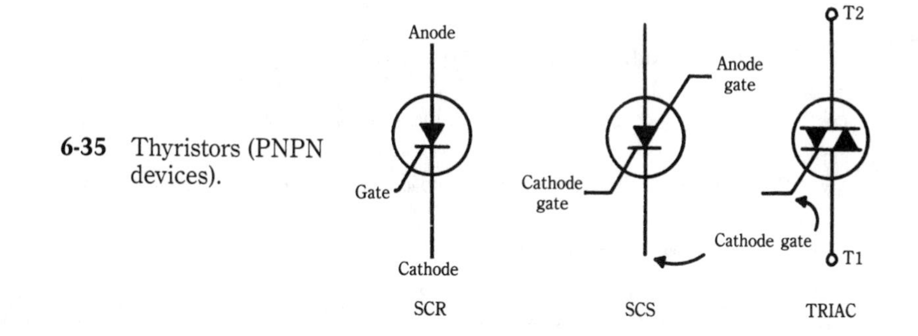

6-35 Thyristors (PNPN devices).

The *silicon-controlled switch* (SCS) is designed so that it can be turned on with either polarity of gate voltage. Like the SCR, reverse voltage from anode to cathode will not result in current regardless of gate pulses. The *triac*, however, can be turned on in either direction with either polarity of gate pulse.

A diac symbol is shown in Fig. 6-36. The diac has only two junctions like a transistor. Therefore, it is not a thyristor. The diac stays a high impedance with either polarity of terminal voltages until a critical voltage is reached. Beyond this voltage the diac goes into an avalanche mode and becomes a low impedance. The answer to Q14 is c.

6-36 Diac.

Integrated circuits

Many of the transistor and diode devices can be incorporated on a silicon chip. When many devices are combined to form complex circuitry on a single chip, the result is called an *integrated circuit* or IC for short. There are many package varieties of ICs including 8, 10, and 12 lead TO-5 "can" types. Dual in-line packages or DIPs are quite common in 8 lead, 14 lead, 16 lead, and many more leads (for larger packages).

Many varieties of linear integrated circuits are now available. Complete audio i-f sections, for AM, FM, and TV are available in IC packages. Chroma demodulators, chroma subcarrier, stereo multiplex decoders, audio amplifiers, and many other IC chips are available. Operational amplifiers are also available in many IC varieties.

Additional reading

Douglas-Young, John: *Technicians Guide to Microelectronics*, Parker Publishing Co., 1978.

Hoft, Richard G., ed.: *SCR Applications Handbook*, International Rectifier Corp., Semiconductor Division, 1974.

Horn, Delton T.: *Basic Electronics Theory—with projects and experiments*, TAB Books, 1981.

Jones, Thomas H.: *Electronics Components Handbook*, Reston Publishing, 1978.

Larson, Boyd: *Transistor Fundamentals and Troubleshooting*, Prentice-Hall, 1974.

Veley, Victor F.: *Semiconductors to Electronics Communications Made Easy*, TAB Books, 1982.

Warring, R.H.: *Understanding Electronics—2nd Edition*, TAB Books, 1984.

7
Basic circuits

Quiz

Below are circuits you should be familiar with:

Q1. The circuit in Fig. 7-1 is:

a. a bridge rectifier circuit.
b. a dc regulator.
c. a dc amplifier.
d. a 130-volt dc power supply.

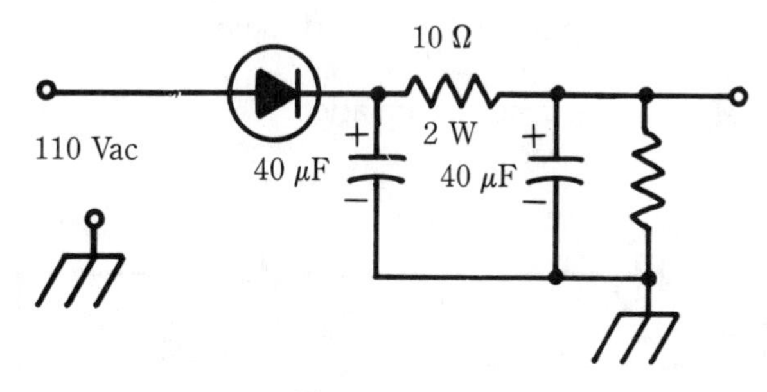

Figure 7-1

Q2. The circuit in Fig. 7-2 is:

a. a brute-force power supply.
b. a video detector circuit.
c. an audio detector circuit.
d. a 130-volt power detector circuit.

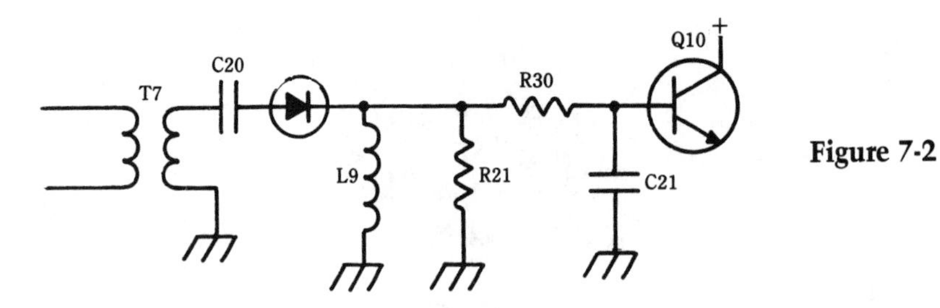

Figure 7-2

Q3. The circuit in Fig. 7-3 is:

a. a transmitter power output stage.
b. a Colpitts oscillator.
c. a crystal oscillator.
d. a one-shot oscillator.

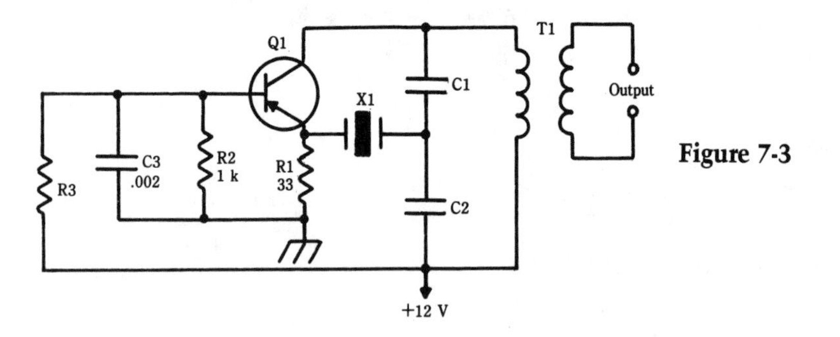

Figure 7-3

Q4. The circuit in Fig. 7-4 would be used to:

a. supply negative dc power supply voltage at the output.
b. serve as a frequency doubler in an FM radio stereo receiver.
c. detect video in a color TV.
d. detect FM radio frequency information.

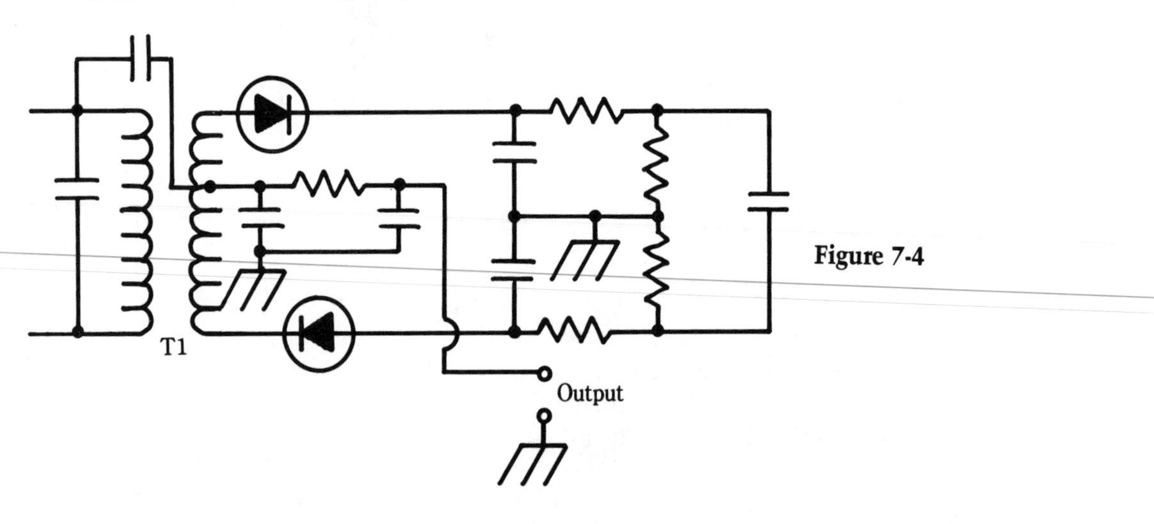

Figure 7-4

Q5. The circuit in Fig. 7-4 also can be called a discriminator.

a. True
b. False

Q6. The device in Fig. 7-5 is:

a. an opto-isolator.
b. an LED segment driver.
c. a photo resistor.
d. a photo transistor.

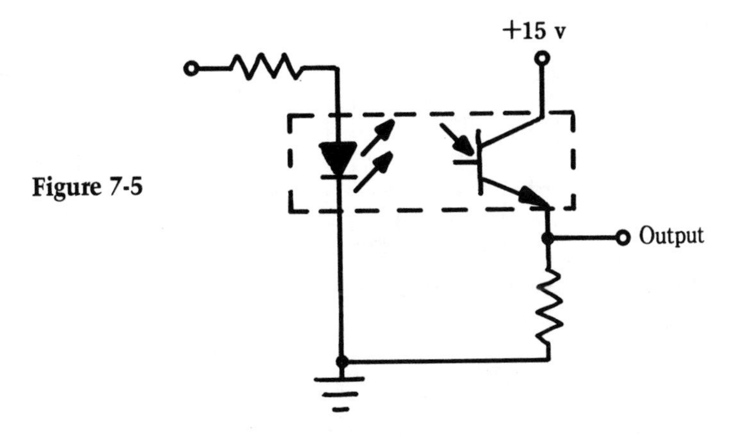

Figure 7-5

Q7. The circuit in Fig. 7-6 is:

a. a transistor flip-flop.
b. an astable oscillator.
c. a multivibrator.
d. all of the above.

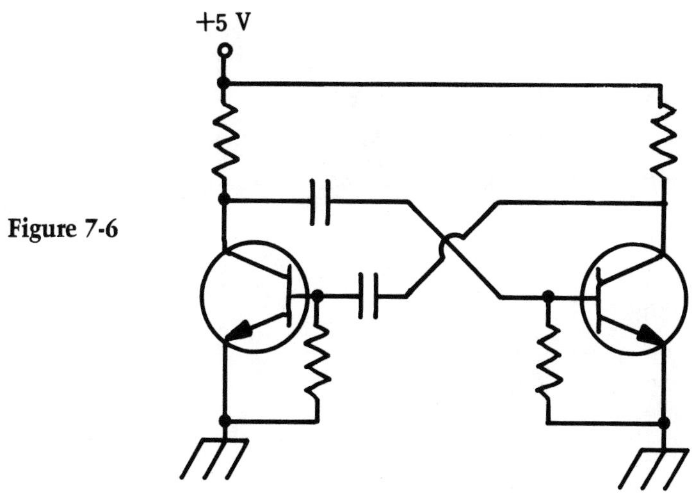

Figure 7-6

Q8. The circuit in Fig. 7-7 is:

a. a discriminator.
b. a ratio detector.
c. a dc power supply.
d. a color TV burst amplifier.

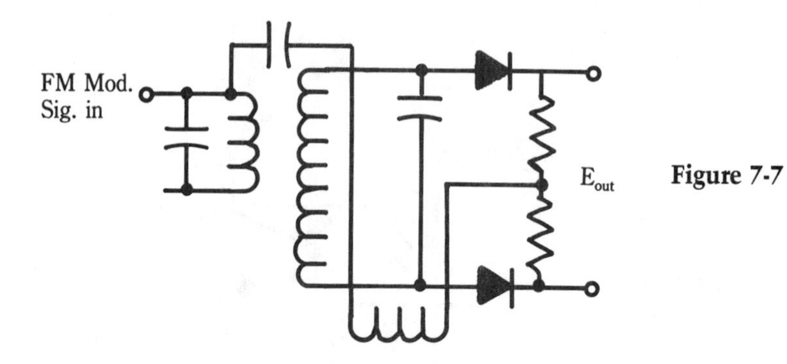

Figure 7-7

Q9. The circuit in Fig. 7-8 is:

a. a power supply regulator.
b. a complementary audio output circuit.
c. a Darlington-pair audio output stage.
d. a push-pull audio output stage.

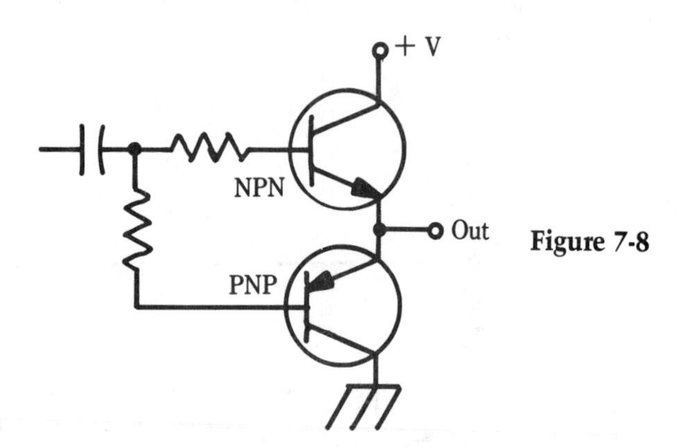

Figure 7-8

Q10. The circuit in Fig. 7-9 is:

a. a voltage comparator.
b. a signal amplifier.
c. a stereo audio amplifier.
d. a J-K flip-flop.

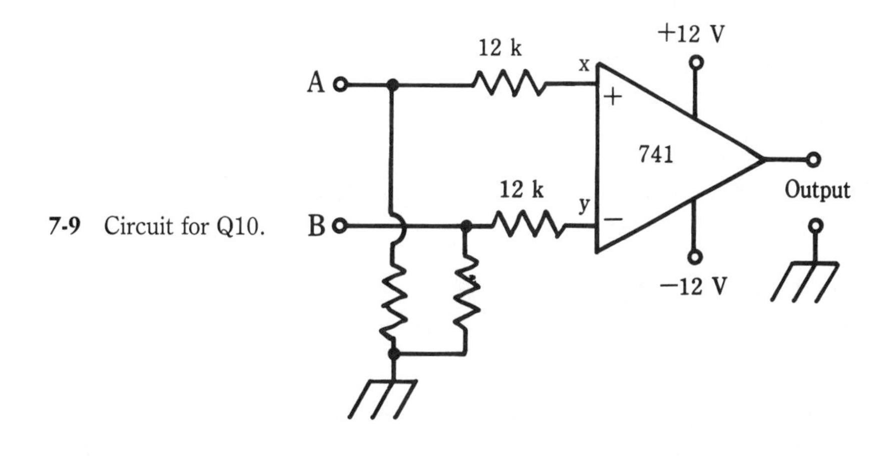

7-9 Circuit for Q10.

Q11. The circuit in Fig. 7-10 is:

a. a bridge rectifier circuit.
b. a passive regulator.
c. an active regulator.
d. an audio amplifier.

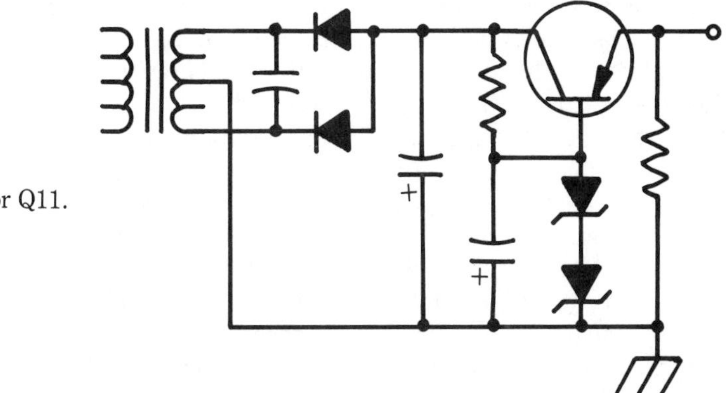

7-10 Circuit for Q11.

Recognizing electronics circuits

There are many other circuits that you should be able to recognize in your work as a technician. Here are some you might already know:

Figure 7-1 is a common power supply circuit as used for decades in table radios and TV sets. It is called a *brute-force* half-wave rectifier because no transformer is required and the 110-V ac is coupled directly to the rectifier diode D1. Originally this type of circuit used a vacuum-tube diode for D1. Later selenium multiplate rectifiers were used in place of the tube. Then silicon diodes were seen as an improvement, and that is what you will find in most cases today. The 110-volt ac sine wave is rectified, meaning only 1/2 cycle causes the rectifier to conduct. This charges C2 up to the peak (1.414 times 110 volts = about 150 volts) input value of the ac sine wave. R1 and C3 are part of the "pi" filter that smooths out the dc sufficient to power the vacuum tubes or transistors in the product. Typical values of C1, C2, and C3 might be 50 to 150 microfarads. R1 might be 10 to 100 ohms. This type of circuit required a *floating ground* so that the chassis could not be connected directly to the ac line.

Figure 7-2 looks somewhat similar to Fig. 7-1. It has a transformer input (T-1), which could conceivably be connected to the power line. But here, the output goes to the control element of the transistor. Therefore it must be a "signal" that is being processed—not power being produced. Obviously the choke coil L1 would short any dc voltage to ground. The coupling tranformer T1, has undoubtably some bandpass characteristics limiting the incoming signals to those desired. D1 rectifies the AM frequencies allowing the original modulation of the AM signal envelope to be derived. R2 and C2 filter out any remaining RF. With that explanation, can Fig. 7-2 be anything other than an AM detector? No. However, because TV video is an AM signal, there is a likelihood that it is a video detector, rather than audio. Both answers would be accepted on the CET exam.

Looking at Fig. 7-3, the first clue telling you this is an oscillator is that there is no input to Q1— only an output. Next, notice the inclusion of X1, a crystal. Crystals vibrate when a voltage is impressed on them. They are selected, or *cut*, to a thickness that causes them to vibrate at desired known frequencies. When a crystal is the determining factor regarding the oscillator's output frequency, the circuit is called a *crystal oscillator*.

There are several identifiable configurations for oscillators. Figure 7-6 is a type of oscillator called a *multivibrator*. It produces a square-wave output, or flip-flop signal, whereas the crystal oscillator typically produces a sine-wave output.

Armstrong, Colpitts, and Hartley are three types of oscillators well known from decades of use in early radio and electronics technology. The Armstrong oscillator's recognizable feature is the use of a *tickler* coil. This is a unique winding within the feedback transformer that accomplishes the task of coupling some of the output signal back to the input to initiate and maintain oscillation. The tickler coil is recognizable because it is a winding within the small transformer that is terminated within—only one wire connects to that winding. While this tickler is a coil, or inductor, it can be thought of as coupling by way of the transformer's interwinding capacitance.

The Colpitts is recognizable as having no tickler coil and no crystal. In its grid or base circuit, it has dual split capacitors that provide the proper feedback from output to input to sustain oscillations near the desired frequency.

The Hartley oscillator is recognizable because it has none of the previous three features; instead it has split inductances in the grid or base circuit through which feedback is derived from output back to the input. This feedback kicks the circuit into oscillation and continues to make the oscillator run at its desired frequency.

The multivibrator mentioned earlier (Fig. 7-6) has been used in television circuits as the horizontal oscillator.

The vertical oscillator in TV sets often used what was called a *VBO* or *vertical blocking oscillator*. It was distinguishable because at the very low frequencies in that circuit (60 Hz), the feedback transformer used was necessarily rather large in size (1 inch in diameter is common). This iron-core transformer coupled the large plate or collector amplified 60 Hz signal back to the grid or base circuit. A variable resistor (vertical hold control) varied the base or grid bias and RC time to correct the oscillator frequency to near that needed. The vertical sync pulse then kicked the circuit into conduction at exactly the right time to be in sync with the transmitted picture frame or vertical sync pulse.

Electron-coupled oscillators are the most difficult to recognize. However, with no input to the circuit drawing you should suspect an oscillator is the function. ECOs are able to oscillate, whether a vacuum tube is used or a transistor, by utilizing the inter-electrode or inter-element capacitance inherent in any electronic component. The problem is that this capacitance is not shown on most schematics, making the circuit difficult to classify. Fortunately these are not widely used today.

Today, oscillators are more often *VCO,* or *voltage controlled oscillators*, and *PLL,* or *phase-locked-loop* oscillators.

The VCO can be a multivibrator as in Fig. 7-11 and Fig. 7-12. The frequency of the flip-flop action can be controlled by varying a dc voltage to the base of one transistor, which causes the speed of oscillation to increase or decrease while maintaining equal duty-cycles for both the positive and negative portions of the square-wave output.

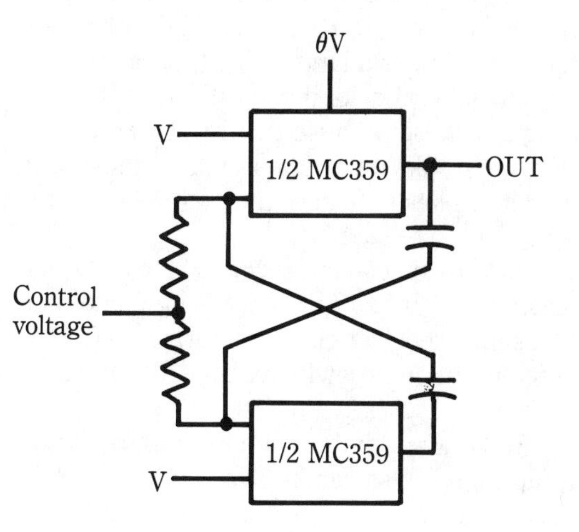

7-11 IC multivibrator circuit.

Phase-locked-loop systems use a detector to determine whether the oscillator used in the radio or other device is drifting off frequency. According to the direction that the frequency attempts to drift off, up or down, the detector produces a dc control voltage, + or −. This control voltage is fed back to the oscillator, thus maintaining the exact desired oscillator frequency.

Figure 7-4 is an FM ratio detector used to extract the audio modulation from an FM signal. Because it is similar to Fig. 7-7 (another FM detector), you need a

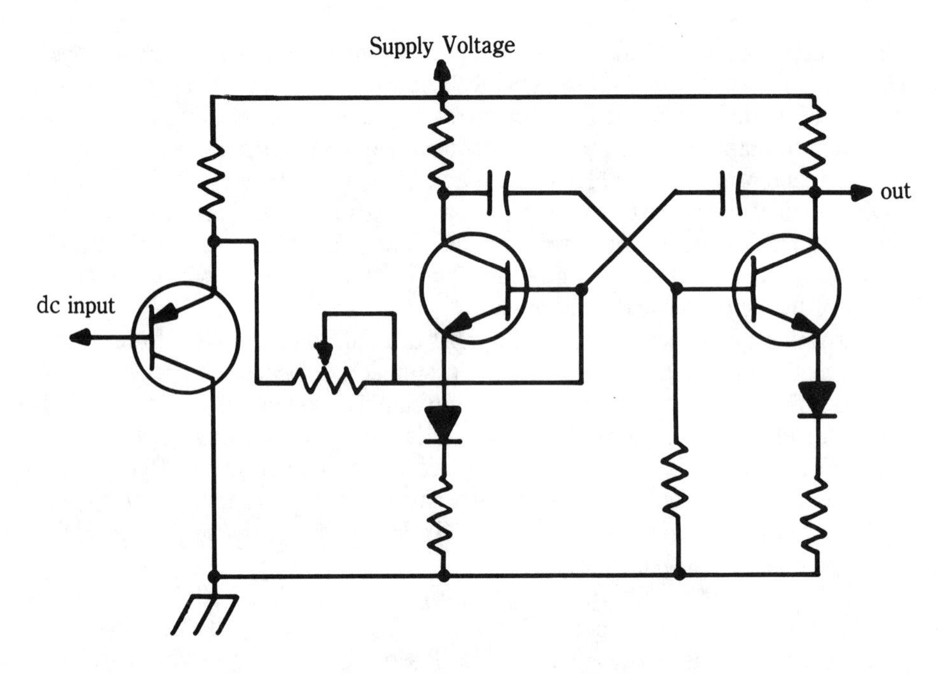

7-12 Discrete multivibrator circuit.

means of telling them apart. The key to recognizing a ratio detector is to note that the two diodes are reversed. The output is derived from the center of the transformer and uses a large value (perhaps 5 μF) capacitor in the output.

Question 5 asks if the ratio detector also can be called a discriminator. The answer is false because they are two different types of circuits. The discriminator (Fig. 7-7) looks much like the ratio detector in Fig. 7-4. Looking carefully, you will note its diodes both have their cathodes on the output. Also there is no large capacitor.

In Q6 a special but recognizable component is shown. It is commonly called an *opto-isolator*. It has a light emitting diode that might be energized by a remote control signal receiver circuit to turn on a TV set. Such a circuit eliminated the early mechanical on-off switches on TV sets and instead operated a relay or solid state switch.

In Q7 a multivibrator, a form of oscillator, is shown. It is also commonly called a flip-flop. It also can be named an astable oscillator (not bi-stable). Therefore the correct answer to Q7 is d, a better answer than either a, b, or c.

Figure 7-8 is an audio output circuit. The object of this circuit is to allow a higher level of power than would be available using only a single transistor of the same general size and cost. These complementary transistors are usually selected to have similar parameters and are sold in sets of two. Notice Fig. 7-8 allows power to be delivered to the output (the circuit containing a speaker coil) without the expense or infidelity of a power output tranformer.

Figure 7-9 is an op-amp. Operational Amplifiers are used a lot in electronics; an entire book could be written on op-amps and their uses and circuit configurations. In this case the op-amp is used as a signal comparator. It senses small changes in either of the inputs, A or B. A slight change in either input drives the output either full positive, or full negative.

Figure 7-10 is a full-wave dc power supply utilizing an active regulator circuit. Without the transistor and zener diodes, the full-wave rectifier circuit would produce a negative dc voltage. By using a pi filter made up of very large filter capacitors and either a large inductor or energy wasting resistor, the output ripple will be reduced. A better method, using smaller capacitors and being more efficient, is the active filter as shown. Increases in load current or input voltage, within limits, are closely controlled by this circuit.

Figures 7-11 and 7-12 are examples of VCOs, or Voltage Controlled Oscillators. While these flip-flop oscillators will free run, they are more useful in many cases, if their frequency can be precise. Various types of feedback circuits are used in conjuction with these VCOs to provide locked frequency settings.

Phase-locked loop allows radios to be tuned using a potentiometer, which varies the voltage to the oscillator. It also is used with now common chip ICs for detection circuits. Most of these circuits are named on the schematics you will find yourself using in troubleshooting.

8
Decibels

Editor's note: Chapter 8 is a direct reprint of the material from chapter 7 of TAB Book No. 2910, *The CET Communications Exam Book*, by Dick Glass and Ron Crow.

Decibel notation is used in several areas of communication electronics. This chapter reviews this important system.

Quiz

Q8-1. Decibel notation can be based on particular values, such as dBs referred to one millivolt.

a. True
b. False

Q8-2. Basing the dB notation on 1 millivolt, then 2 millivolts would be equal to 0 dBmV:

a. True
b. False

Q8-3. If 10,000 microvolts is equal to 20 dBmV, then how many dBmV is 20,000 microvolts?

a. 10,000 dB
b. 20,000 dB
c. 2 dB
d. 26 dB

Q8-4. When referring to the output levels of an amplifier versus the input level, or an antenna's forward gain versus its rearward signal pickup, dBs are used to indicate:

a. current gain.
b. the ratio between two quantities.

c. voltage drops.
d. power supply input.

Q8-5. Ordinarily, in master antenna systems, it is easier to calculate the line, splitter, tapoff and insertion losses by adding, subtracting, multiplying and dividing the microvolt levels, rather than adding or subtracting one- or two-digit dB expressions:

a. True
b. False

There are few areas in electronics that don't eventually end up using decibels. Especially the area of antennas and signal distribution systems, or audio technology. Technicians should be able to easily handle decibels or quickly understand quantities expressed in dBs. The problem is that dBs aren't common-usage terms (like inches, pounds, or gallons) which we all can easily visualize. It's hard to visualize 6 dBs. How big? How wide? How many are 6 dBs? On the other hand, dBs really aren't difficult if a little time is spent getting used to them.

A few concepts can clear it up

With a few basic truths in mind, the calculation of the values and levels used with dBs can be handled easily.

Two ways

One problem in understanding dBs is that we use them in two different ways: In TV we establish 0 dB as 1000 microvolts (μV). When discussing signal strength in TV and MATV systems, then, we can communicate by stating a 100,000-μV signal is also a 40-dB signal. Or, a 500-μV TV signal is a −6-dB signal. So dBs are used to indicate a "specific voltage level" for any dB amount. This is convenient if you are frequently dealing with signals that all relate to one base, or set amount of signal. See Table 8-1.

0 dBmV

Why is 1000 μV important? In earlier design of TV sets it was determined that a TV receiver should be manufactured so that it would produce a "snow free" video picture with an input signal of 1000 μV. This then became a convenient reference and was set at 0 dBmV. TV set makers couldn't design one set to operate on 50 μV and another on 1 full volt. So 1000 μV, or 1 millivolt, was the design standard. If you remember that a TV set requires 1000 μV, or 0 dBmV of signal, you will already know the most important single fact you need to understand dBs. When speaking of TV signal levels from an antenna, or in an MATV system, you can refer to any signal level as it relates to 0 dBmV. Any technician who works on antennas for long will know that a −20-dBmV signal level is going to produce a very snowy picture and a +20-dBmV signal will not only be ample to supply a perfect noise-free picture, but could be split easily to supply 2, 3, or 4 more TVs and all would have snow-free picture quality.

Table 8-1. Chart of dB vs. microvolts.

Micro-volts	dBmV	5-Place dBmV	Micro-volts	dBmV	5-Place dBmV
10	-40	-40.00000	1995	6	6.00000
16	-36	-35.91760	2000	6	6.02060
32	-30	-29.89700	3200	10	10.10300
63	-24	-24.01319	4000	12	12.04120
100	-20	-20.00000	8000	18	18.06180
125	-18	-18.06180	10000	20	20.00000
250	-12	-12.04120	16000	24	24.08240
316	-10	-10.00000	32000	30	30.10299
500	-6	-6.02060	63000	36	35.98681
560	-5	-5.03624	64000	36	36.13260
630	-4	-4.01319	100000	40	40.00000
700	-3	-3.09803	127000	42	42.07607
800	-2	-1.93820	250000	48	47.95880
891	-1	-1.00250	320000	50	50.10300
1000	0	0.00000	500000	54	53.97940
1100	1	.82785	1000000	60	60.00000
1300	2	2.27887	2000000	66	66.02060
1400	3	2.92256	4000000	72	72.04120
1600	4	4.08240	8000000	78	78.06180
1800	5	5.10545	10000000	80	80.00000

Expressing a signal level relative to 0 dBmV is a custom we have adopted in television-radio work. But because dBs are used in comparing human hearing and in audio work, 0 dB sometimes means something else.

The other way

The other way to use dBs in electronics is in expressing the *difference* between one signal level and another. An easy way of visualizing this is to compare the amount of signal you would expect from two different antennas—one stronger than the other. See Fig. 8-1.

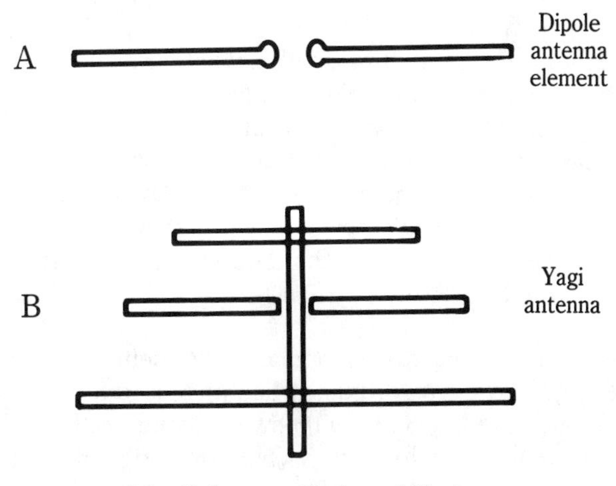

8-1 Reference dipole and Yagi.

One antenna, A, is a basic 1/2-wave dipole. Other antennas are all based on, or compared with it. The antenna in A, no matter what amount of microvolts are produced at its terminals, is said to have 0 dB of signal. It is the standard starting point. Other antennas don't actually have gain in the sense that they amplify the signal received. Stating that an antenna has a 6-dB gain means that it will pick up more signal (twice as much as a simple dipole). It can't be predicted that one antenna will have an exact microvolt level of signal at its terminals, because the antenna could be sitting directly under a broadcast antenna, thus receiving "volts" of signal, or it might be a hundred miles from the antenna and lucky to receive 50 μV. However, knowing that an antenna with 6-dB gain will perform twice as well as a basic 1/2 dipole gives something to compare with and to visualize.

A second example would be a need to supply a second TV set from an existing antenna, as well as the main set. To divide the signal a splitter is needed that properly maintains the impedance of the system (75 ohms when using TV coax). All signal splitters basically have the same loss, 3.5 dB. Splitters do not divide a signal in half. They are better than that in that they only lose about a third of the signal to each output of the 2-way splitter. Knowing that the splitter loses 3.5 dB means you can accurately determine what signal will be available at the TV sets. It doesn't mean that the microvolt level at the TV terminals is known, unless the input signal level is known. It does mean that component and wire losses, regardless of what the input signal level is, can be determined.

In the above example, the signal was split to two televisions rather than one. If an antenna was supplying only −10 dBmV of signal, and another 3.5 dB was lost through the splitter, only −13.5 dB would be left at the TV sets. Or, if +6 dBmV was supplied by the antenna and the splitter loss was 3.5 dB, there would still be +2.5 dBmV at the TV sets. The −13.5-dBmV signal is going to be intolerable. On the other hand, the +2.5-dBmV signal left over from the hypothetical +6-dBmV signal is still excellent.

It's important to know that these wire losses, splitter losses, and tapoff losses are something that can be relied on. To determine a system's losses you don't need to know how much signal a TV set must have; you don't need to know what the signal level is coming off an antenna, or out of an amplifier. That's good. No matter what the losses, they remain the same, in dBs, whether dealing with a small microvolt signal or volts of signal. This makes it quite easy to figure MATV systems or to determine what signal will be available after you install an antenna and run coaxial cable to three or four rooms in a home. By using dBs to calculate gain or loss in signal systems, the numbers are small, usually only one or two digits. If microvolt levels were used it would be confusing with millionths of a volt and hundreds of thousands of microvolts and other large and small numbers.

6 dBs

Six dBs is the secret to using dB mathematics. Remember that. A gain of 6 dB means whatever signal level you originally had increased to twice as much. A signal level of 1 volt, increased by 6 dB, will now be 2 volts. A 200-μV level increased by 6 dB will now be 400 μV. No matter what your original level, a 6-dB level is twice the original.

Minus 6 dB is 1/2 the original. A 1-volt signal to begin with, when reduced by 6 dB, is 0.5 volt (or 500 millivolts). The 200-μV level, if reduced by 6 dB is now only 100 μV. You can calculate signal levels anywhere in a system by simply halving or doubling voltage levels.

Let's use another example to show how easy it is. Start with 5 volts of signal. After amplification you now have 25 volts. What amount of gain did you produce within the amplifier? Obviously it is a gain of 5. However to use dBs to express that gain let's first double 5 volts to 10 volts. That is a 6-dB gain. Now double the 10 volts to 20 volts. That's another 6-dB gain. Now we have 12 dB of gain. Let's double the 20 volts to 40 volts. That would be another 6-dB gain. But wait. We wanted only 25 volts, not 40. Reviewing, 20 volts was 12 dB of gain. 40 volts is 18 dBs. So, 25 volts is somewhere between 12 and 18 dBs of gain. In antenna work, knowing that might be all you need to know to solve a problem or design a system. But I'm sure you can come closer to the actual dB gain the 25 volts is than either 12 or 18 dBs. We know it isn't half the way, or distance between 20 and 40 microvolts, which would be 30 μV. So 25 would probably not be a gain of 12 + 3, or 15. It would be less than 15 dB of gain. Let's guess that it would be 14 dB. If you guessed 15, that would be close enough for practically all system calculations you will meet up with. We can scientifically calculate it to an exact dB amount, because we know that

$$\text{dB} = 20 \log \frac{V_2}{V_1} = 20 \log 5 = 13.979$$

Another factor is involved in dB loss calculations. Loss in a TV system will be much less for channel 2 frequencies than for channel 83. Wire loss will be over twice as much for channel 83. Therefore, to demand that you be precise in loss calculations accomplishes little. The fact that you don't have to be exact is not a reason for you to remain cloudy in using dBs. Think of it as making your job easier and less demanding.

A chart based on dBmV

Look at Table 8-1. Start by first assuring yourself that 0 dB is listed as 1000 μV. It is. Therefore you know all the dB levels shown will be referenced to 0 dB, the basic starting point for RF signals. That makes it easy. Now you and I could discuss a signal problem or we could design a signal distribution system noting the signal levels at any point. We would both know what each was saying to the other.

Analyzing the table further, notice that it only goes negative to −40 dBmV. Actually that 10-μV signal is not negative. It is truly a +10-μV signal. It is only negative when compared with our standard starting point, 0 dB.

Why stop at −40 dBmV? Anything lower serves no purpose. A 10-μV signal is so small, it is virtually no signal. The FCC rules governing CATV systems require that any stray signal leakage on the cable system can measure at no more than −40 dBmV. That is just about zero signal so far as a TV antenna or TV set is concerned. When a far off TV station appears faintly on a normally unused channel, you see a barely discernible signal, usually unable to lock, with so much snow that you cannot recognize objects in the picture, nonexistent color and hissing sound—that is a −40-dBmV signal.

On the other end of the scale the chart shows 80 dBmV or 10 volts of signal. That is such a giant signal that most TV sets would be swamped. Preamps and line amplifiers cannot handle such a large signal input. Ordinarily in your work, you will find 35 to 40 dBs of signal is too much for most systems and too large of a signal for a TV set. TVs work great on −6 dbmV to +25 dBmV. Anything less is snowy. Anything more and adjacent channel interference becomes a problem. 80 dBmV is frequently used in large CATV and cable system trunk lines.

It isn't perfect

To illustrate that the use of dBs does not ordinarily require exactness in calculations, in Table 8-1 compare 1 dBmV with 2 dBmV. The chart shows a 200-μV difference in the levels. Notice the difference between 2 dBmV and 3 dBmV. The chart shows only a 100-μV difference. Actually 2 dBmV is a little less than 1300 μV, and 3 dBmV is a little more than 1400. But it's close enough for most of the work we do.

Picking the specks out of the pepper

When you finally understand how useful dBs can be, you might need to be more exact; here is a formula for dB calculation when dealing with voltage levels (not power levels).

$$\text{dB} = 20 \log \frac{V_2}{V_1}$$

Suppose you have a voltage level of 200 μV that is to be amplified to 1300 μV. Let's do it the easy way first:

200 doubled = 400 μV or 6 dB
400 doubled = 800 μV or 12 dB
800 doubled = 1600 μV or 18 dB

So 1300 is about 60 percent of the difference between 800 and 1600. Let's guess 1300 μV represents 16-dB gain over 200 μV.

Now let's use logs:

$$\text{dB} = 20 \log \frac{V_2}{V_1}$$

$$\text{dB} = 20 \log \frac{1300}{200}$$

$$\text{dB} = 20 \log 6.5$$

$$\text{dB} = 20\ (0.81291)$$ (Taken from a log book.)

$$\text{dB} = 16.26$$

Obviously you have to use a logarithm chart to calculate the gain using the log formula. You rarely carry one in your pocket. A lot of us would get confused with the algebra anyway. So we goofed in the dead-reckoning when we guessed 16 dB might be the gain, rather than 16.26 dB. But the guess was close enough for most antenna work.

Watts confusing?

Now that you have all the above down pat and you really feel you can handle dBs, let's change the rules! When calculating watts, the rule changes. Now the dB formula is:

$$\mathrm{dB} = 10\log\frac{P_2}{P_1}$$

It is not:

$$\mathrm{dB} = 20\log\frac{V_2}{V_1}$$

This means a gain of 6 dB is not a doubling of the power as it was with dBs that dealt strictly with voltage. A 6-dB increase is a quadrupling of the original power. A 3-dB increase is a doubling of the power.

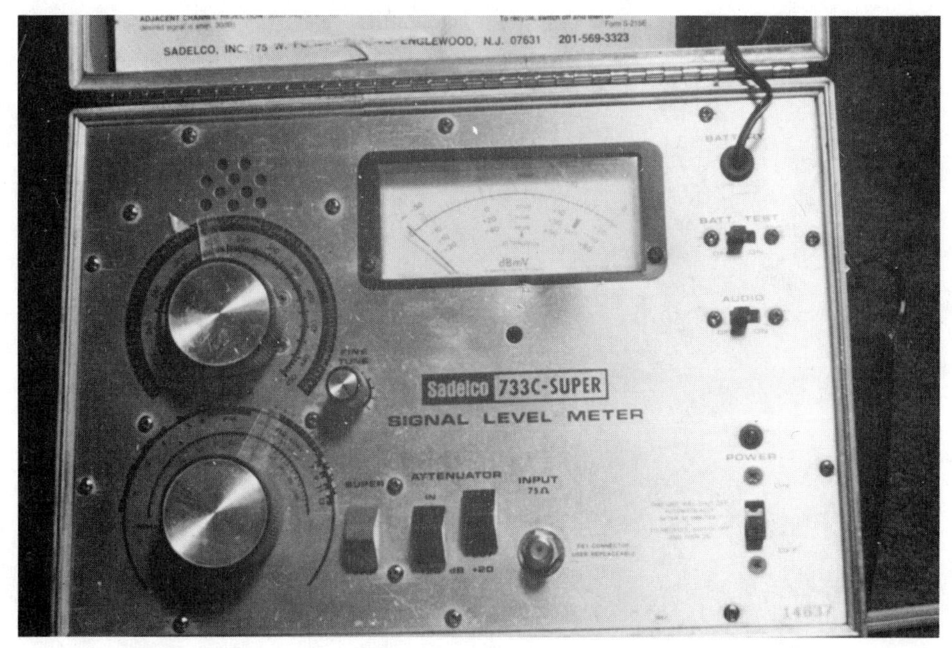

8-2 Television frequencies signal-level-meter measures dB levels.

Perhaps you are confused. Why change the rules of the game, right when you thought you had it? Well, really the rules haven't changed. The problem is Ohm's law! A watt is made up of both volts and current. Here is the power formula:

$P = EI$, or Power = Volts times Current

An example would be:

$P = EI$

$P = 2$ volts times 2 amps = 4 watts

Now double both the voltage and current.

$P = 4$ volts times 4 amps = 16 watts

You can see that doubling both the voltage and current would not double the power. Instead it multiplies it by 4.

So, in your work with power levels, or audio levels, or your own hearing level, a power increase of +3 dB is twice the original level. A −3-dB loss is 1/2 the original (Fig. 8-2).

9

Antennas and wave propagation

Quiz

Q1. In antenna systems 0 dB is:

a. total lack of signal.
b. less than 1 volt peak-to-peak signal.
c. 1000 microvolts signal level.
d. insignificant noise level within the desired signal.

Q2. Figure 9-1 represents what type of antenna?

a. dipole
b. whip
c. yagi
d. all-channel TV antenna

Q3. Impressing RF signal voltages on the elements of an antenna such as in Fig. 9-1 will cause the transmission of those RF signal voltages in an amount directly related to the power input.

a. True
b. False

Q4. If Fig. 9-1 is a bird's-eye view of a transmitting antenna, in what direction will maximum RF signal be propagated?

a. A
b. B
c. C

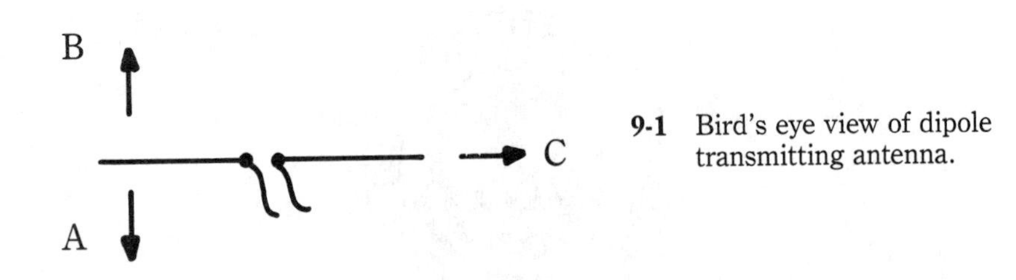

9-1 Bird's eye view of dipole transmitting antenna.

Q5. An antenna like that in Fig. 9-1 will respond to those frequencies that are shorter in wavelength than the resonant frequency of the antenna, but will be much less efficient at lower frequencies with longer wavelengths.

a. True
b. False

Q6. At resonance a half-wave dipole antenna typically has what impedance?

a. 50 ohms
b. 72 ohms
c. 300 ohms
d. 1000 ohms

Q7. The speed of wave propagation from an antenna is the same as the speed of light.

a. True
b. False

Q8. The speed of light is:

a. 5280 miles per second.
b. 250,000 miles per second.
c. 186,000 miles per second.
d. 186,000,000 miles per second.

Q9. The wavelength of a 1-hertz signal would be:

a. 984 million feet.
b. 5280 feet.
c. 5280 miles.
d. 186,000 miles.

Q10. The wavelength of a 1-megahertz signal would be:

a. 984 feet.
b. 492 feet.
c. 246 feet.

Q11. The wavelength of an RF signal of 100 megahertz is:

a. 9.84 feet.
b. 4.92 feet.
c. 2.46 feet.

Q12. The wavelength of an RF signal of 1000 megahertz (1 gigahertz) is:

a. 0.984 feet, or nearly 12 inches.
b. 0.492 feet, or nearly 6 inches.
c. 2.46 feet, or nearly 2.5 inches.

Q13. If the formula for wavelength is

$$\frac{984{,}000}{\text{f (in kHz)}}$$

What is the wavelength of a CB radio signal of 27 megahertz?

a. about 4 feet.
b. slightly over 32 inches.
c. 9 feet.
d. 36 feet.

Q14. The speed of propagation in the metal of an antenna, or the transmission line between a receiver or transmitter, is lower than that of free space.

a. True
b. False

Q15. In Fig. 9-2, which drawing best depicts a coax cable's loss elements as they appear to a UHF signal?

a. A
b. B
c. C

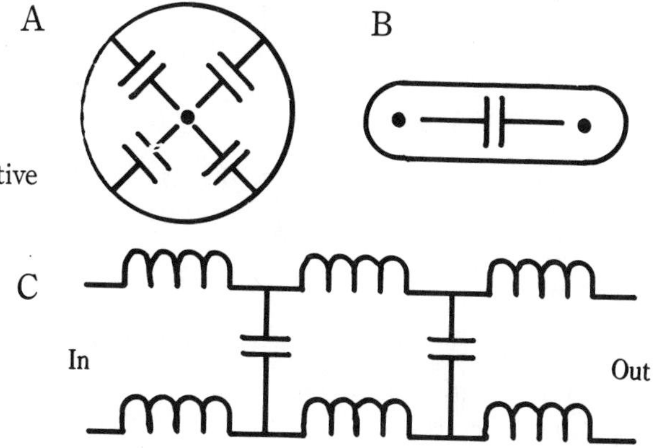

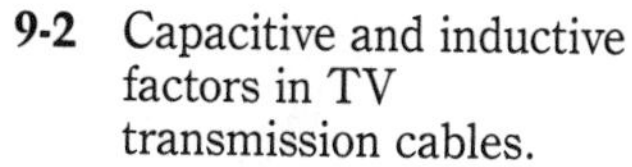

9-2 Capacitive and inductive factors in TV transmission cables.

Q16. The polar pattern, or graph, of an antenna's ability to propagate an RF signal will be as follows:

a. Vertical dipole is Fig. 9-3.
b. Horizontal dipole is Fig. 9-4.

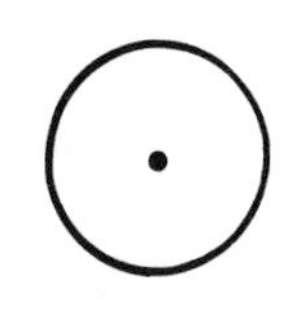

9-3 Horizontal and vertical dipole polar patterns.

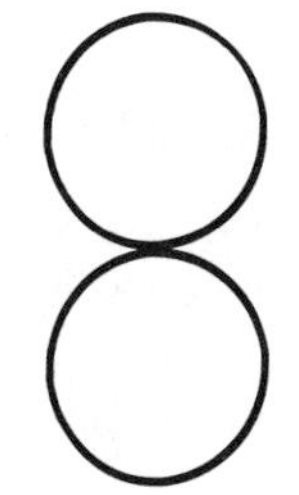

Figure 9-4

Q17. Which of the following patterns in Fig. 9-5 depicts CW transmissions?

a. A
b. B

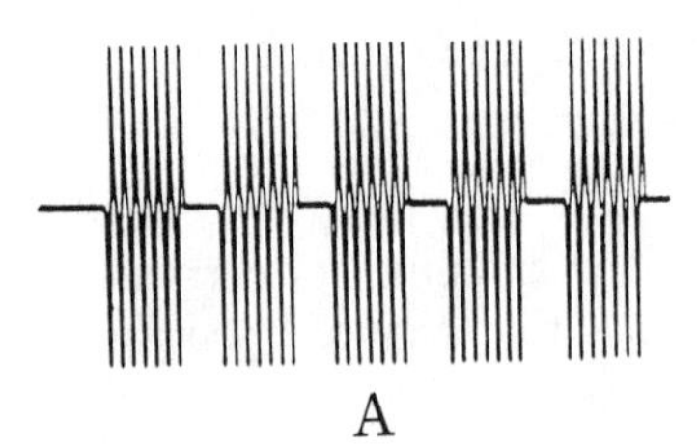

9-5 Transmission waveforms.

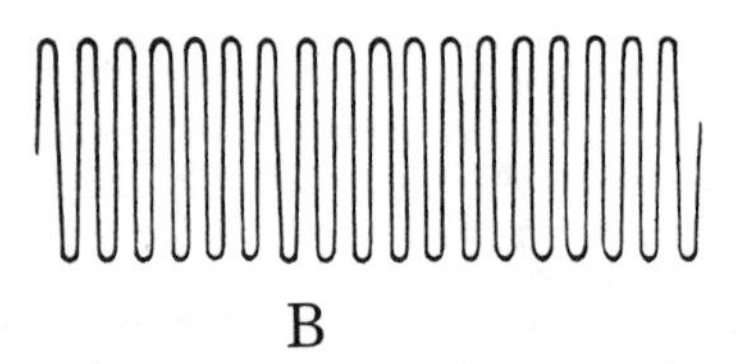

Q18. Which of the patterns above might have been useful in transmitting Morse Code?

a. A
b. B

Q19. Would the waveform in Fig. 9-6 be called CW, or Modulated?

a. CW (continuous wave)
b. modulated

9-6 The frequency is the same, the amplitude varies.

Quiz explanation

Because the previous chapter was devoted entirely to the subject of dBs, Q1 should be easy. The answer is c.

Question 2 asks what type of antenna is depicted in Fig. 9-1. This is the basic antenna against which others are compared, the half-wave dipole.

Not so long ago it was discovered that a piece of metal would produce electromagnetic and electrostatic waves, if impressed with an RF voltage. Higher amounts of voltage cause stronger waves to be emitted. The answer to Q3 is true.

Question 4 asks if you understand that the direction of propagation of the signal from the dipole in Fig. 9-1 is at right angles to the antenna element, not out the ends. There is no single correct answer because A is just as correct as B.

Question 5 wants to know if you realize that an antenna works best at its resonant frequency. The resonant frequency of a half-wave dipole antenna is slightly shorter (about 95 percent) than the half wavelength of the signal. While it might work to some degree at those frequencies near its resonant frequency, its efficiency will drop off the further the frequency is either higher or lower than that of resonance.

Question 6 asks if you know the impedance of a half-wave dipole at resonance. It is 72 ohms. The other impedance you need to remember is that of a folded dipole. Its impedance is 300 ohms.

Question 7 asks if you understand that electromagnetic waves all travel at the speed of light in space. Whether the transmission is at medium-high frequencies, very-high, or super-high, the speed is the same as light. Light itself is an electromagnetic wave in a frequency spectrum above that used by humans for most broadcast services. When working with UHF signals and C Band satellite signals, the similarities with light waves increase.

Question 8 asks if you know the frequency of radio wave transmissions. The feet in a terrestrial mile is 5280. The approximate distance to the moon is 250,000 miles. The speed of radio and light waves in space is 186,242 miles per second. The correct answer is c.

Question 9 follows Q8 by asking you to determine how long the wavelength of a 1-cycle-per-second electromagnetic wave would be. If you got Q8 right, then the answer to Q9 must be that the wavelength of a 1-Hz signal is 186,000 miles.

Question 10 is more realistic. You must understand how long a 1-MHz wavelength is. A wavelength of a millionth-of-a-second cycle would be a millionth of 186,000 miles-per-second (as in Q9), or 0.186,000 mile (approximately 982 feet). A 2-MHz signal wavelength then is about 490 feet; 4 MHz is 245; 8 MHz is 122; 16 MHz is 61 and 32 MHz is about 30 feet.

Question 11 requires wavelength calculation. Wavelength for 100 MHz is much shorter than that of 30-MHz. Using the formula for wavelength:

$$\text{Wavelength} = \frac{984}{\text{f (MHz)}}$$

you find that the wavelength is 9.84 feet. Is that correct? If it is, then a half-wave dipole resonant to that frequency should be half of 9.84, or about 4.92 feet long. It is a little shorter because signals travel a little slower in metal. This frequency (100 MHz) is located in the middle of the FM broadcast band. Would an antenna a little longer than 4 feet seem reasonable for an FM antenna? Or for the TV channels on either side of the FM band, channel 6 or 7? So our answer must be close to being right.

Question 12 continues to familiarize you with the lengths of antennas at some of today's broadcast frequencies. 1000 MHz is 110 MHz above the TV UHF band upper channels. The wavelength at this 1 GHz is 0.984, so an antenna element at half wave would be something over 0.4 feet, or around 5 inches. The answer to Q12 is b.

Question 13 requires you to figure out what a CB radio antenna length is. CB antennas on autos are about 4 feet in length, so you could guess that answer. But if you use the formula for wavelength, the wavelength is really about 36 feet. How can this be? Well, first realize the longest CB antenna would be a half-wave dipole less than 18 feet long. Then consider that the auto CB antenna is really a quarter wavelength. That brings the size down to less than 9 feet. Isn't that about what the old bumper CB antennas were? To shorten the most popular styles, the antenna maker used a coil in the base of the antenna, which effectively and electronically allows the antenna to be even shorter than 9 feet. Common sizes are 3, 4, and 5 feet long. The correct answer to the problem in Q13 though is b.

Question 14, as discussed, is true.

To answer Q15 you must understand that as frequency increases the transmission cable begins to look more and more like series inductances and perpendicular capacitances to the signal. Think of coax being less like a shield and center conductor, and more like diagram C in Fig. 9-2.

The understanding of the polar pattern of an antenna is required for Q16. A vertical diople transmits in 360 degrees—in all directions perpendicular to the single metal whip or dipole. Therefore the pattern is like Fig. 9-3. The horizontally polarized dipole, as in Fig. 9-4 can't emit in the direction the dipole elements are pointed. It can transmit in the direction perpendicular to the elements. Thus the polar pattern is a figure 8, as in Fig. 9-4. (Both a and b are correct.)

You must understand Continuous Wave transmission for Q17. It looks like Fig. 9-5. To use CW, you need to interrupt the CW to make the dots and dashes used in Morse Code. A better method would be to use CW, but to modulate the CW with a tone that corresponds to the time a code key is energized by the radio operator. So Q18 requires the b pattern as the correct answer. The a answer has no information contained in that CW.

The CW with modulation impressed on it is shown in Q19. The answer is b.

Figure 9-7 shows a television waveform. It is an example of more than one type of modulation in a single waveform. From the left it has two or three cycles of video

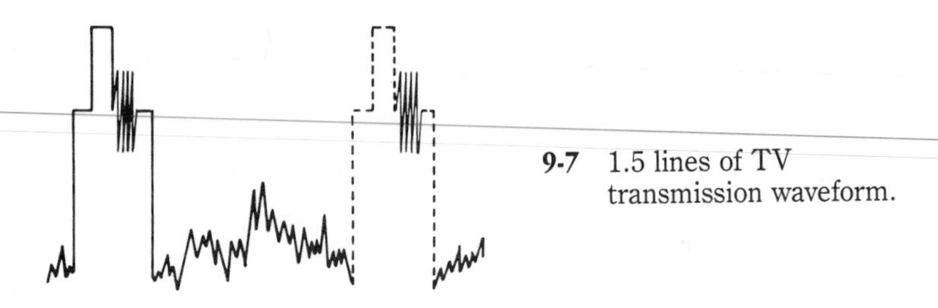

9-7 1.5 lines of TV transmission waveform.

modulation. More of this is contained between the two horizontal blanking pedestals. The highest squared-off portion of the pedestal is the horizontal sync pulse. It is an example of pulse modulation. On the "back porch" of the blanking pedestal are few cycles of CW, or color-burst information.

10
Block diagrams

Quiz

Q1. After ascertaining the symptoms, what should you do first in troubleshooting a seemingly bad radio receiver?

a. Measure power supply voltages.
b. Check speaker voice coil resistance.
c. Inject RF signal at mixer input.
d. Check for signal at the detector.

Q2. If the power supply voltages in Fig. 10-1 are normal, what might you do next to find the cause of the malfunction?

a. Divide the block diagram in two by either injecting an audio signal after the detector, or checking for a signal into the detector.
b. Check speaker terminals for audio signal.
c. Check oscillator for output frequency.
d. Check output signal at audio output stage.

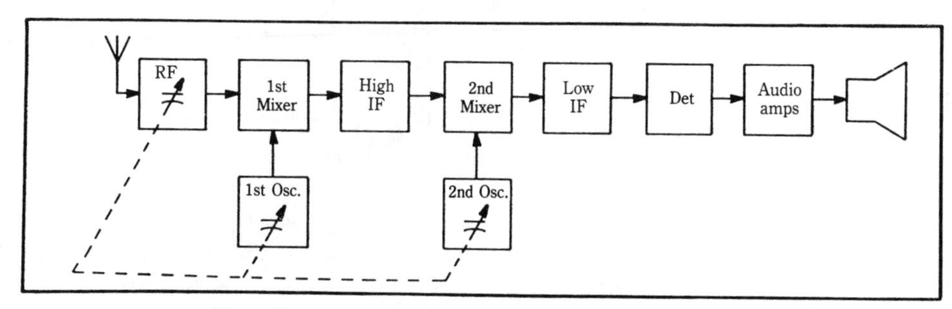

10-1 Block diagram of double-conversion receiver.

Q3. If the audio signal injected at the volume control produces proper sound from the speaker, and the oscilloscope RF probe shows a proper IF signal, what can be the reason for the dead radio symptom?

a. antenna open
b. oscillator dead
c. mixer inoperative
d. detector inoperative

Q4. In the block diagram of Fig. 10-2, block X is:

a. TV receiver circuitry
b. oscilloscope circuitry
c. radar circuitry
d. FM radio circuitry

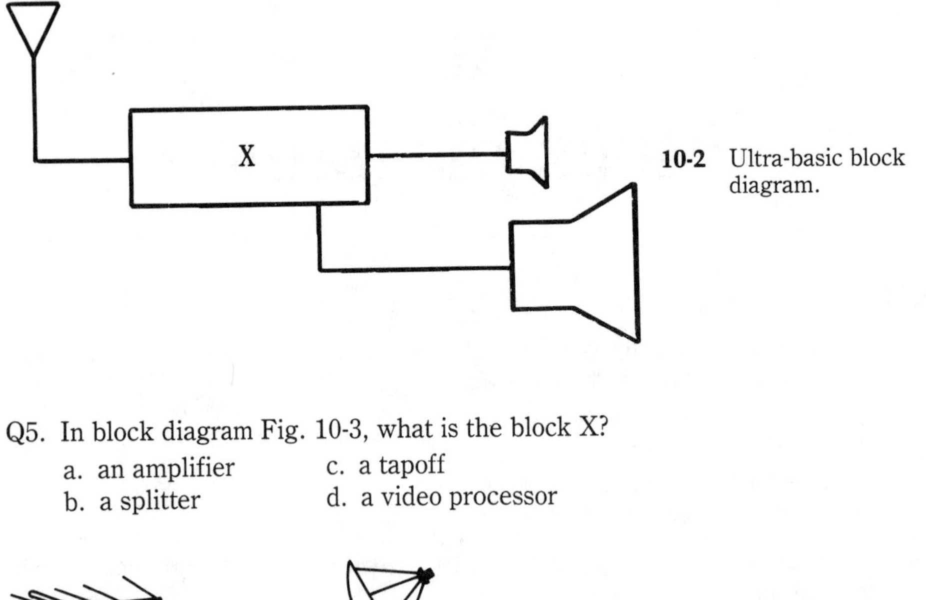

10-2 Ultra-basic block diagram.

Q5. In block diagram Fig. 10-3, what is the block X?

a. an amplifier
b. a splitter
c. a tapoff
d. a video processor

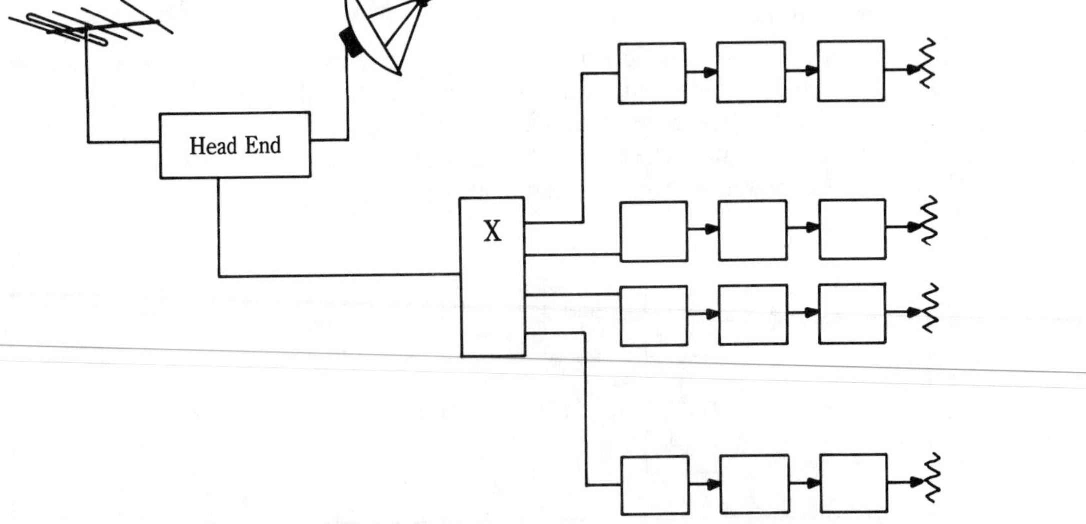

10-3 SMATV (Satellite-Master Antenna TV) system block diagram.

Q6. In the block diagram of Fig. 10-4, what significant block is not shown?

a. agc c. detector
b. audio d. tuner

Q7. Having only a horizontal white line across the center of the TV screen, using the block diagram in Fig. 10-4, what block would the problem most likely be in, of those listed?

a. E c. F
b. CRT d. G

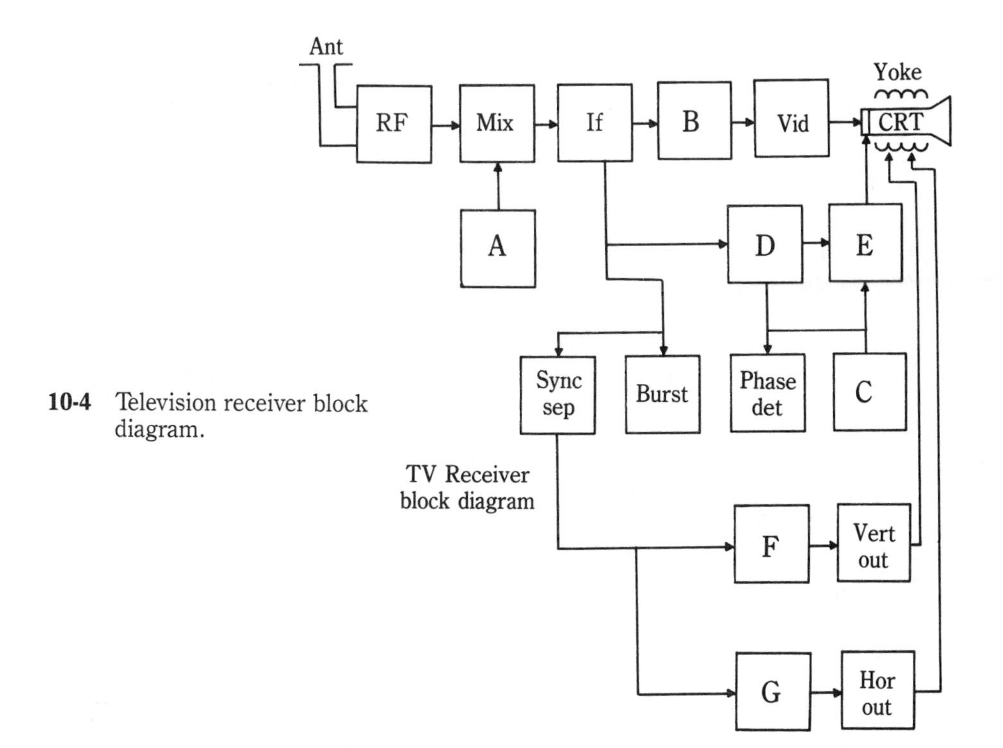

10-4 Television receiver block diagram.

Q8. The block diagram of an audio chip in Fig. 10-5 shows:

a. an intermediate frequency input is needed.
b. it is made for AM radio reception.
c. it is made for FM radio reception.
d. a volume control will connect to C.

Q9. A rectifier diode bridge would be located where in the regulator-power supply diagram Fig. 10-6?

a. A c. C
b. B d. D

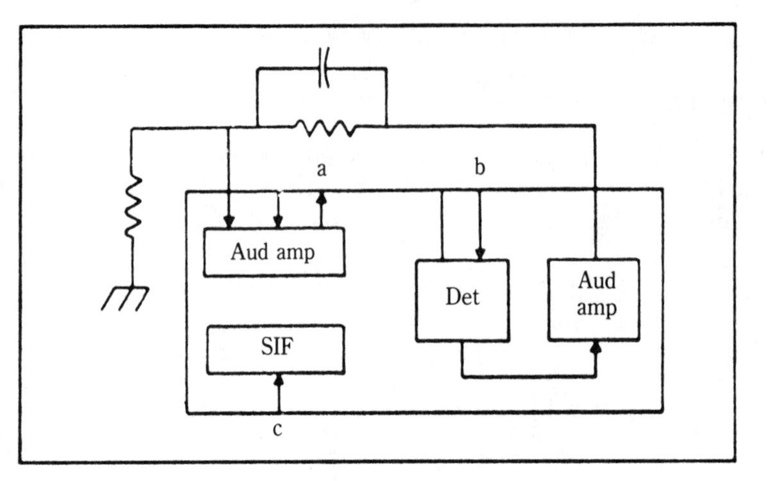

10-5 Diagram for Q8.

Q10. In Fig. 10-6 a series voltage regulator transistor will be located where?

a. A c. C
b. B d. D

Q11. In Fig. 10-6 a reference voltage zener diode would likely be in which block?

a. A c. C
b. B d. D

Q12. If Fig. 10-7 is the actual circuit represented by Fig. 10-6, in which block would you expect to find Q1?

a. A c. C
b. B d. D

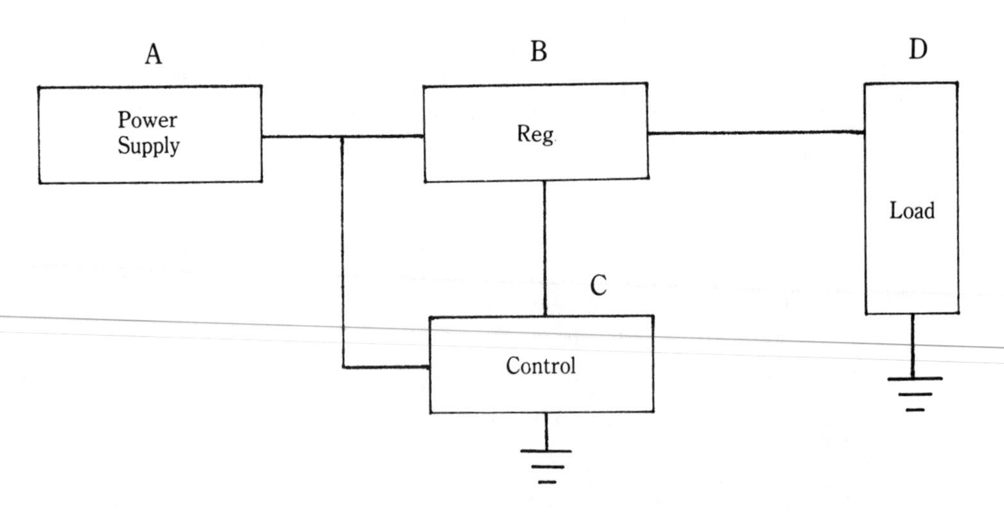

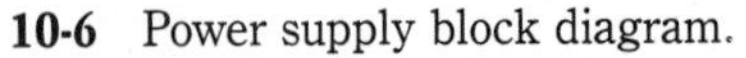

10-6 Power supply block diagram.

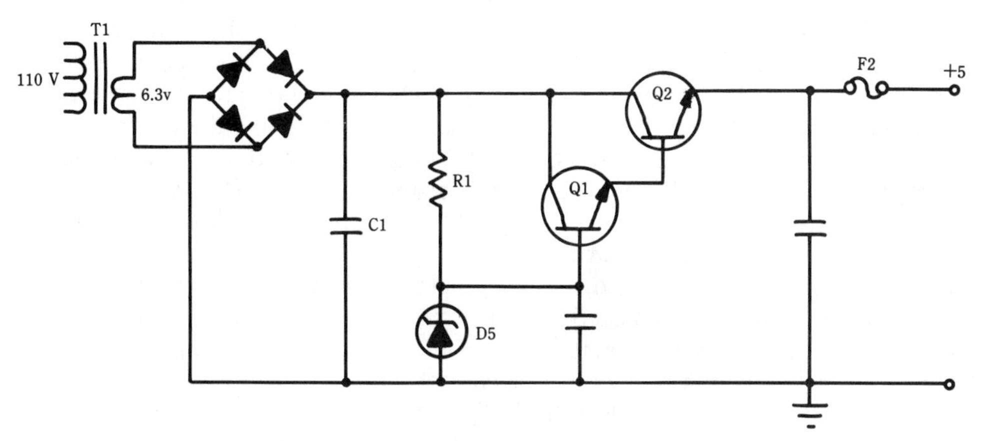

10-7 Circuit of block diagram Fig. 10-6.

Quiz explanation

Block diagrams are used in electronic servicing to explain the functions of the individual circuits. They also show the direction of signal flow, control, or process. They should also show how key sections such as data, signal, or power are connected. The purpose is to give the technician an understanding of the operation the product is supposed to perform. Once the technician understands how each section fits into the total operation, he/she can "divide and conquer" in searching out malfunctions. Once the defective block is identified, circuit troubleshooting takes place with voltage, current, and resistance checks, signal insertion or oscilloscope readings. These narrow the problem down to individual components such as chips, transistors, caps, inductors, resistors, other parts and wiring. Often, electronic design intermingles various components and wiring, which can make troubleshooting or circuit analysis time consuming. Block diagrams help unravel the maze and make sense of the circuitry.

In Q1, looking at Fig. 10-1, a basic AM radio block diagram is shown. It could be more detailed; for instance, it could use three blocks for the IF amplifiers, if there were three separate stages of intermediate frequency amplification. Perhaps the audio section (amp) should be separated to show the first audio amp and the audio power output amplifier section. Block diagrams can be more detailed, or less detailed. Figure 10-1 is sufficient for most of us. In Q1, the question is somewhat arbitrary. If your job is to service hundreds of the same type of radio, rather than dividing it in two to begin to narrow the trouble down, you might have found through experience that a dead radio usually is caused by a specific weak element, perhaps the speaker. In this case you would always look for signal at the voice coil first, which could make your efforts more efficient.

This question asks you about ordinary troubleshooting procedures, where you don't have the advantage of working on the same model with routine fixes. Before performing signal injection or tackling individual circuits, a dead unit's power supply should be checked to assure the supply voltages are proper. A is the better answer.

In Q2, after ascertaining that the power supply is OK, you should attempt to divide and conquer. Near the middle of the circuitry is the audio detector. It is a convenient place to divide the radio and find out whether the trouble is in the front end or in the audio section. A is the best answer. Simply touching the volume control terminals with your finger often gives you a quick check of the audio amps and speaker. You could use an RF probe and oscilloscope to see if the RF has made it to the input of the detector diode. This would eliminate the front end of the radio as the culprit. A is the better answer.

Question 3 states that the audio section appears OK. Also the proper IF signal appears at the output of the last IF. What is the logical trouble? First check the detector diode itself, because you have eliminated just about everything in front of it as well as everything in back of it. D is correct.

Question 4 shows the simplest block diagram. In this case the major components are shown. A detailed block diagram of the receiver (X) is not shown. This diagram would be useful, perhaps, to an installation technician, but not a board-level troubleshooting technician. Many present-day electronics products makers publish block diagrams no more detailed than that of Fig. 10-2, expecting their field service personnel to ship the major sections back to the maker for component level repairs.

Question 5 shows how block diagrams are most useful in master antenna (MATV) and cable TV (CATV) work. While the head-end and block X are similar in size on this block diagram, X is only a simple 4-way signal splitter. The head-end is perhaps a room filled with thousands of dollars worth of receivers, modulators, amplifiers, and connectors. B is correct.

Figure 10-4 is the subject of Q6. It shows a block diagram of a common TV receiver. Notice that the audio detector, amplifier, and output (now on a single chip in many cases) are not shown. B is correct.

Question 7 asks you to use Fig. 10-4 and localize the problem, knowing that there is no vertical deflection. Go to block F to see if it is producing a vertical pulse. If so, check the vertical output block. If it is proper, then it could be that the vertical yoke coils around the neck of the cathode ray tube are open. C is correct.

Question 8 shows in Fig. 10-5 a more modern block diagram, this of a single chip audio circuit. Looking at the functions within the chip, you can determine that C is no place to connect a volume control. You can't determine that it is for AM or FM. You can see the sound IF signal inputs at C. The chip is probably used in a TV receiver. A is the best answer.

The next three questions ask what portions of the power supply/regulator circuit contain the major components. Figure 10-7 is the actual circuit, showing the diode rectifier bridge and power transformer that would be represented by block A. Most schematics show the remainder of the circuit as the regulator. Today's electronic products more frequently use a single regulator 3-pin chip that contains all of the regulator and control components except the capacitors. In that case the block diagram should show the block A, B, and D only.

Knowledge of the operation of this regulator is mandatory for technicians. Confined, small-profile electronics products build up excess heat that results in stress on all of the power supply components. Overstressed, the diodes short, the capacitors dry up, the zener shorts, or the pass and control transistors fail. A great percentage of technician repair time is spent on these regulator circuits. Computers, TVs, radios, and satellite receivers commonly have 2, 3, 4, or more of these circuits, each supplying a different dc voltage for operation of various portions of the product circuitry.

Regulator operation

The power transformer T1, having about a 20:1 turns ratio, drops the 120 Vac line voltage to 6.3 volts. The common full-wave bridge circuit rectifies the 60 hertz ac, outputting across C1 a dc voltage near 7 volts (remember the 6.3 volts is RMS, not peak). This dc contains a small amount of ripple at 120 hertz. Question 2 is a silicon current amplifier able to pass current of up to 0.5 amp. It has the ability to dissipate up to 20 watts of collector power.

By carefully controlling the base voltage on Q2, the 5-volts regulated output is maintained at an exact dc level. Should the load current through F2 increase, the 5 volts at the Q2 emitter would tend to drop. However, the drop causes Q2 to increase conduction because the drop at the emitter increases the b/e bias. If the relative voltage on the base is more positive than it previously was, relative to the emitter, the transistor conducts more, keeping the 5 volts at the desired 5-volt level.

If the bridge diodes produce a lower rectified dc output, perhaps because the T1 transformer received a lower ac line voltage input, the dc supplied to the top of C1 and R1 (as well as to the transistor collectors) will be lower. Instead of 7 volts, let's say it is now only 6.9 volts. This lower voltage also shows up at the cathode of the zener, D5, causing it to conduct less, producing a lower IR drop across R1. This means the base of the NPN transistor Q1 is more positive. This causes the emitter to conduct more raising the base voltage on Q2, causing increased conduction of Q2 and thus maintaining the 5-Vdc output to the load at F2. The exact same thing happens during any peaks and valleys remaining at the bridge rectifier output. That's why we no longer listen for hum in a radio speaker to determine if the power supply and output circuits are working, like we did in the 30s, 40s, and 50s. The ripple is virtually eliminated by the regulator circuit.

The TV block diagram

Refer again to Fig. 10-4, the TV block diagram; You can see it is for a TV receiver. It is much more complex than that of a radio block diagram. This circuit should be memorized, as all of the TV receivers developed from this basic one. This one is actually very simple compared with modern day TVs that contain remote control, digital on-screen display, picture-in-picture, stereo sound enhancement, autoprogramming, and clock timers. Some are combined with VCRs and radios. If you understand the function of each of the blocks in this drawing, you will have little trouble with expanded TV circuitry.

All technicians should understand TV circuitry. In Fig. 10-4 you can see the top row of blocks receives the TV signals from the antenna. It uses a superheterodyne mixer to derive a single IF frequency, no matter what channel the TV is turned to. The high gain IF supplies the 40 MHz IF frequencies to the detector at block B. The video circuit delays the luminance portion of the TV signal slightly to allow the color signals (which has been processed through the color circuitry) time to catch up. The luminance and color information then reaches the CRT simultaneously to drive the tube properly.

The color information is fed through block D and separated into the proper proportions of red, green, and blue, with the proper intensity of each. This allows the CRT to repaint the picture tube screen with the original camera images. Color control is accomplished by the burst-phase detector-color oscillator at C.

To maintain exact synchronization of the TV picture with the broadcast camera, horizontal and vertical sync pulses are inserted by the camera into the video frames. In the sync separator block of Fig. 10-4, selective circuits separate the sync pulses and send them to blocks F and G, the vertical and horizontal oscillators. The output of the oscillators is shaped and amplified to precisely drive the CRT beams to trace horizontally about 260 times per frame, while moving from top to bottom of the screen once.

TVs are interesting to troubleshoot. If you have a bright screen but no sound or video, you would first inspect the power supply (not shown on the block diagram) for any obvious missing or abnormal voltages. If those appear normal, you should suspect a problem between the antenna and the detector, B. Signal injection, first at the video detector, then the IFs and Mixer might lead to the cause of the problem. The block diagram causes you to attempt to reduce the problem to a single block. In many TVs, the entire block then can be substituted, or an integrated circuit containing multiple components can be further inspected or substituted.

The problem many of us have in electronic troubleshooting is failing to stick to the principle of reducing the problem to a single block. We randomly move around the set first checking the voltages here and there, hoping to run across a glaring cause of the fault, a dead transistor, chip with a wrong voltage, or off-value resistor. In many cases, even with the best intentions, a tech will get lost and end up desperately hoping to stumble onto an elusive defect. Some problems are so difficult you will swear the fault has to be in the bare-metal chassis itself, because you have proved everything else to be in working order. Realizing you can get stumped easily, your best bet is to constantly attempt to divide and conquer; to reduce the possibilities down to one block, then to test the components in that block or substitute it.

11
Digital concepts

THIS CHAPTER COVERS CONCEPTS OF DIGITAL ELECTRONICS PERTINENT to the technician of today. This rapidly expanding field should be understood by almost all electronics technicians. Computer techniques are invading almost every electronics area.

Quiz

The first five questions pertain to the circuit in Fig. 11-1.

Q1. In this circuit if input A and B are both open (no connection), the output voltage will be:

a. +5 volts.
b. nearly zero volts.
c. −5 volts.
d. None of these.

Q2. The base to ground voltage with the same conditions as in Q1 would be:

a. about 0.6 volt.
b. +5 volts.
c. about 0.2 volt.
d. −5 volts.

Q3. If input B is grounded and input A left open, the output would be:

a. +5 volts.
b. nearly zero volts.
c. −5 volts.
d. None of these.

Q4. If positive logic is assumed, the circuit is:

a. a NAND gate.
b. a NOR gate.
c. an AND gate.
d. None of these.

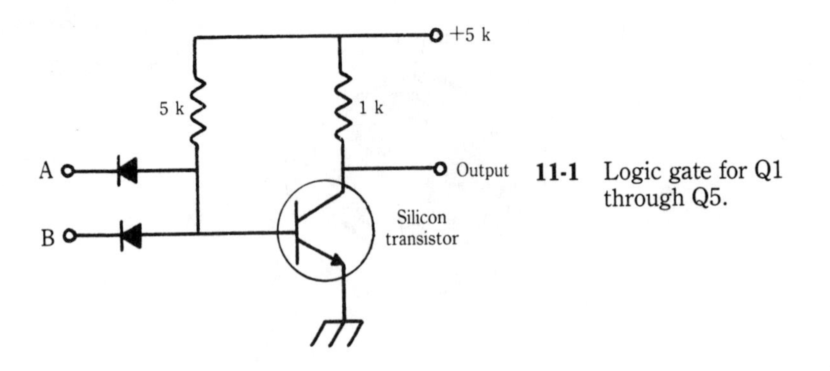

11-1 Logic gate for Q1 through Q5.

Q5. If negative logic is assumed, the circuit is:

a. a NAND gate.
b. a NOR gate.
c. an AND gate.
d. None of these.

Questions 6 through 10 refer to the circuit in Fig. 11-2.

Q6. The signal at G could best be described by the logic equation:

a. $G = BC$
b. $G = CD$
c. $G = A + B$
d. None of these.

Q7. If $A =$ "1" and $B =$ "1" the signal at F will be:

a. logic 1
b. logic 0

Q8. If positive logic is assumed with "1" = +5 volts and "0" = 0 volts and if A, B, and C equal 0 volts, and D equals +5 volts, H must equal:

a. +5 volts.
b. 0 volts.
c. 2.5 volts.
d. None of these.

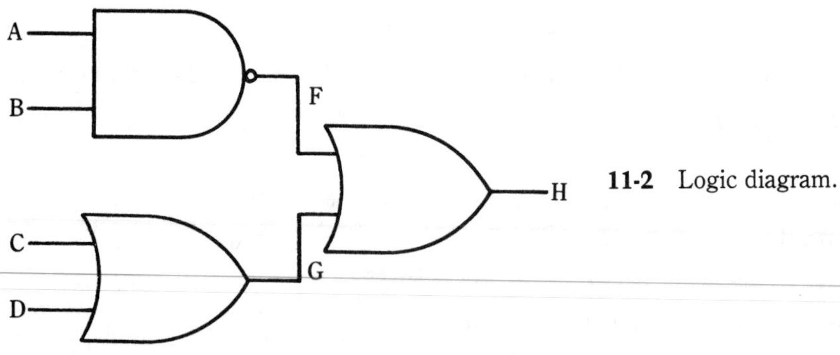

11-2 Logic diagram.

Q9. If A and B are logic 1 and if C and D are both logic 0, H is:

a. logic 1
b. logic 0

Q10. The equation that describes H is:

a. $\overline{A+B}(CD)$
b. $AC+BD$
c. $\overline{AB}+C+D$
d. None of these describe H.

Q11. An SR flip-flop is:

a. monostable
b. astable
c. bistable
d. None of these.

Q12. After the SR flip-flop in Fig. 11-3 is clocked, the output at F will be:

a. +5 volts.
b. nearly zero volts.
c. undeterminable.

Q13. The toggle flip-flop changes state with each trigger:

a. True
b. False

Q14. Data flip-flops are used for:

a. counters.
b. A-to-D converters.
c. tri-state buffers.
d. memory.

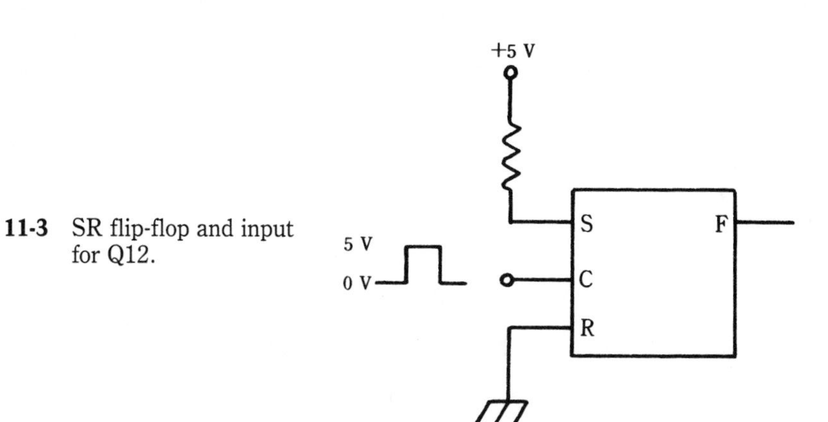

11-3 SR flip-flop and input for Q12.

Switching circuits

At the heart of all computer-digital circuits is the switch. Understanding the transistor switch is, therefore, fundamental to a thorough understanding of digital computers. (Another field of study is involved in understanding computers from a programmer's viewpoint, which is different than an electronics technician's viewpoint.)

In switching circuits little power is consumed in the active device, which is the transistor. The circuit discussion below illustrates why this is the case. When SW1 in Fig. 11-4 is in the A position there is no bias to supply base current for the transistor. The transistor is, therefore, OFF; that is, no collector current flows. With no

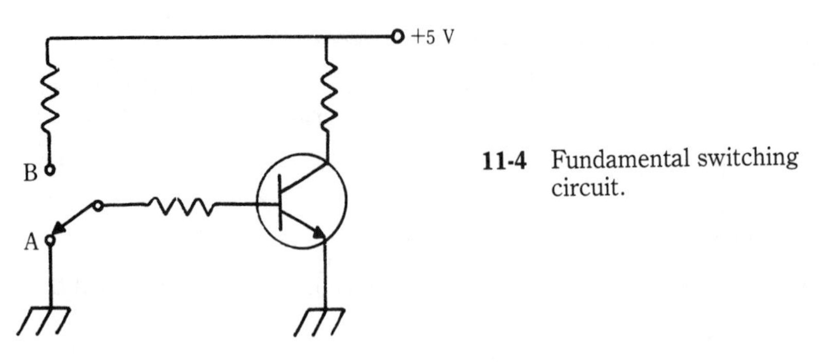

11-4 Fundamental switching circuit.

collector current flowing the collector voltage is +5 volts. The power dissipated in the transistor is:

$$\begin{aligned} P &= E \bullet I \\ &= 5\text{ V} \bullet O_I \\ &= 0 \text{ watts} \end{aligned}$$

(Actually a small amount of leakage current does flow in the transistor. It is in the microampere range.)

When SW1 is in position B the base-emitter junction is forward biased. Resistances are chosen to cause enough current in the base circuit to produce saturation. If the transistor is saturated, the collector voltage is very low, about 0.7 V for silicon. Let's assume a load resistance is in the collector of 1000 ohms. The collector current then is:

$$\frac{5\text{ V} - 0.7\text{ V}}{1000} = 4.3\text{ mA}$$

Then

$$\begin{aligned} P &= (4.3\text{ mA})(0.7\text{ V}) \\ &= 3.01\text{ mW} \end{aligned}$$

Hence, in either the ON or OFF state, the transistor consumes very little power.

Logic gates

This basic transistor switch can be coupled with other circuit components to make various computer gates. Diodes, for example, can be used to form gates and make a useful (if obsolete) logic device when coupled with transistor switches. Let's examine such a circuit in Fig. 11-5.

Assume that inputs A and B are derived from collectors of transistors similar to those in this circuit. In other words, if the collector "driving" A is off, A will be +5 volts. If the collector "driving" A is ON the voltage will be 0.7 volt. The same voltage level might occur at B independent of A. To describe the circuit output at C (the dependent variable in this case), we need to examine all possible combinations of A and B. If A and B are both low (0.7 volt) the diodes cannot conduct. There will be no base current; hence, the transistor is OFF and C is +5 volts. Suppose A is

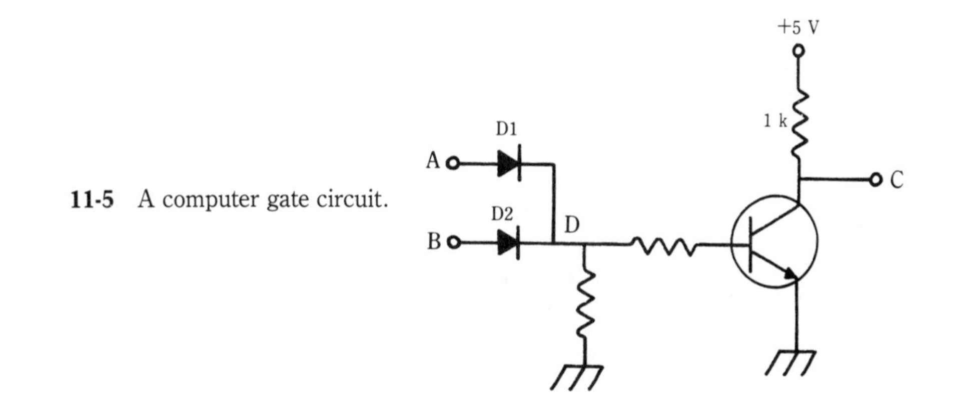

11-5 A computer gate circuit.

high (+5 volts) while *B* remains low. *D*1 becomes forward biased, *D*2 remains OFF. The +5 volts forward biases the base-emitter junction and the transistor goes ON. The voltage at *C* drops to 0.7 volt. In fact, either *A* or *B* high will result in *C* low. *A* and *B* both high will only further drive the transistor on and *C* will remain low. The following table summarizes these statements.

A	*B*	*C*
Low	Low	High
High	Low	Low
Low	High	Low
High	High	Low

In digital circuits we speak of logic levels of "1" and "0." Sometimes true and false are used instead, with true the same as logic 1 and false the same as logic 0. If the numerals one or zero are to be used for logic levels they are set apart by quote marks when used in sentences and not preceded by the word "logic" or the words "logic level."

For positive logic the higher voltage level is considered a "1" or True. A "0" is the lower voltage level. If the circuit above is used in a positive logic system, the following table shows "1" = High and "0" = Low.

A	*B*	*C*
0	0	1
1	0	0
0	1	0
1	1	0

This is now called a truth table because the original work in logic was done using true and false assignments to sentences connected by connectives such as AND, OR, and NOT.

Notice that if this circuit is used in a negative logic system where "1" = low voltage and "0" is the high voltage, the negative logic truth table will be quite different in appearance (even though it's the same circuit).

A	*B*	*C*
1	1	0
0	1	1
1	0	1
0	0	1

If the circuit of Fig. 11-5 is separated into logic functions, the diodes make up a logic gate, while the transistor is an inverter or NOT circuit. For the transistor if the input at *D* is high the output at *C* is low. *C* is said to equal "not *D*" in this case. A bar over a letter is used to indicate NOT. Thus:

$$C = \overline{D}$$

Another way of saying the same thing is in a truth table.

C	*D*
0	1
1	0

Redrawing the gate in the Fig. 11-5 circuit and leaving off the inverter gives the circuit in Fig. 11-6. Notice that in this circuit, the input at *D* goes high when either *A* or *B* goes high.

A	*B*	*D*
Low	Low	Low
Low	High	High
High	Low	High
High	High	High

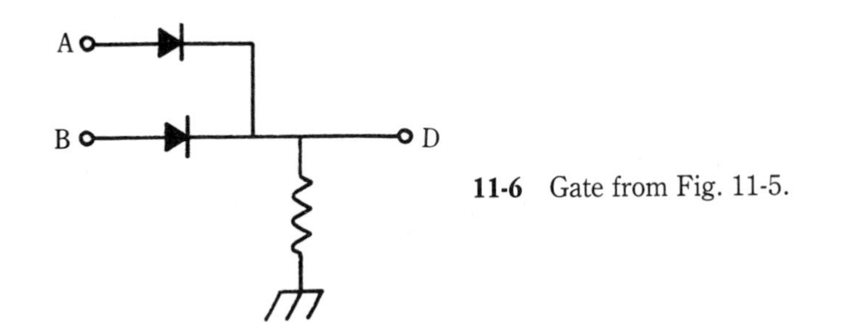

11-6 Gate from Fig. 11-5.

If a positive logic system is involved, the truth table with "1" = high and "0" = low is:

A	*B*	*D*
0	0	0
0	1	1
1	0	1
1	1	1

This means that if either *A* or *B* are "1," the output *D* is "1" this is an OR gate.

If negative logic is the system that contains the circuit of Fig. 11-6, then "1" equals zero or low voltage and "0" equals +5 volts (or high). In this case, the truth table looks quite different. Substituting low = "1" high = "0" in the voltage table gives a truth table that looks like this:

A	*B*	*D*
1	1	1
1	0	0
0	1	0
0	0	0

We get a "1" out only when both *A* and *B* are "1." All other variations of *A* and *B* give a "0" output. Now the circuit is functioning as an *AND* gate.

There are a few logic circuits that use negative logic. Most designers use positive logic circuits. Let's examine the positive logic AND gate in Fig. 11-7. Here, if either or both inputs are ground, the output *D* is low. Only if both inputs are high ("1" for positive logic) do we get a "1" out.

A	*B*	*D*
0	0	0
0	1	0
1	0	0
1	1	1

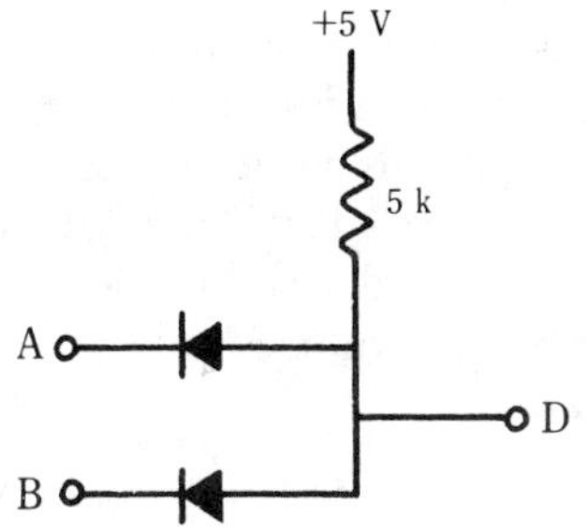

11-7 Positive logic AND gate.

You might want to draw the table for negative logic for this circuit. You should be able to see that it will be an OR logic function for negative logic.

If this circuit is followed by an inverter, it will be the same as the circuit of Q1 to Q5 (Fig. 11-1). When *A* and *B* are left open, the transistor will saturate if the dc current gain of the transistor is 5 or greater. Assuming saturation the output will be nearly 0 volts (actually about 0.7 volt). This is the answer to Q1.

Because the base-emitter is forward biased, the voltage drop across it is 0.6 or 0.7 volt (silicon junction). The answer to Q2 is therefore a.

This circuit is a positive-logic AND gate followed by a NOT circuit. If a truth table is drawn for AND and a column is used to show the inverse of the AND we get:

A	*B*	$A \bullet B$	**Output** $\overline{A \bullet B}$
0	0	0	1

0	1	0	1
1	0	0	1
1	1	1	0

If either A or B is at zero volts (logic 0) the output is "1" or high. (Answer a, Q3.)

The name of this circuit is taken from the combination of the NOT function and the AND function, NOT AND shortened to NAND. (Answer a, Q4.) An OR followed by an inverter is a NOR. The truth table for the NOR is:

A	B	$\overline{A+B}$
0	0	1
0	1	0
1	0	0
1	1	0

If the above circuit is used in a negative logic system, the positive logic AND will be an OR for negative logic. The OR followed by an inverter will give the NOR function described above in the NOR truth table. The answer to Q5 is, therefore, b.

Logic symbols

The symbols in Fig. 11-8 are used in drawings for the AND gate. The OR gate symbols are shown in Fig. 11-9. These symbols are in pairs. That is, if the second AND gate symbol is used on logic diagrams, the second OR gate symbol also would be used. The Electronics Technicians Association, International prefers the first symbols. A small circle at the input or output line indicates an inverter. Examples of this are shown in Fig. 11-10.

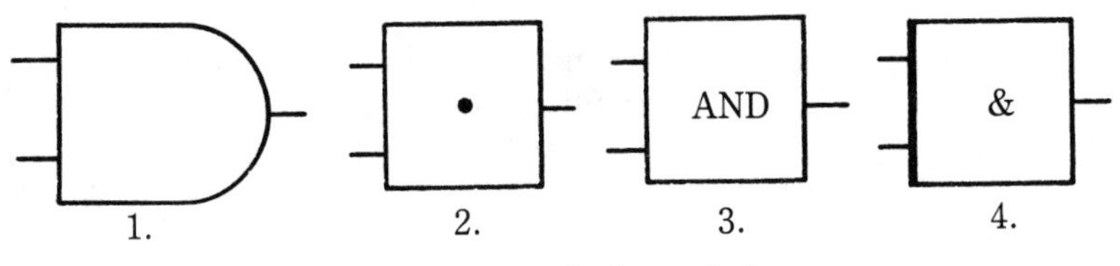

11-8 AND gate logic symbols.

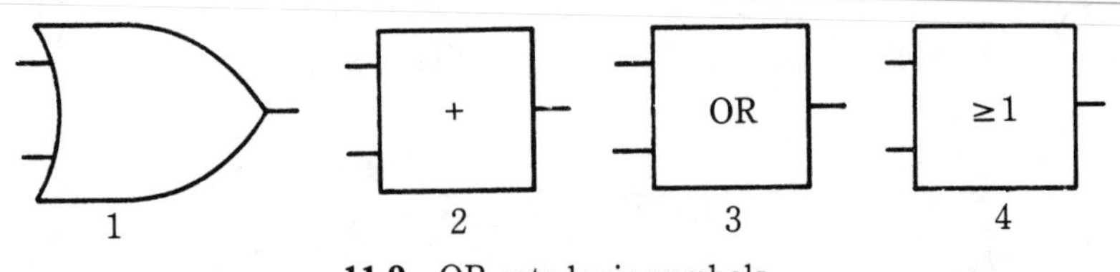

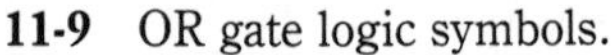

11-9 OR gate logic symbols.

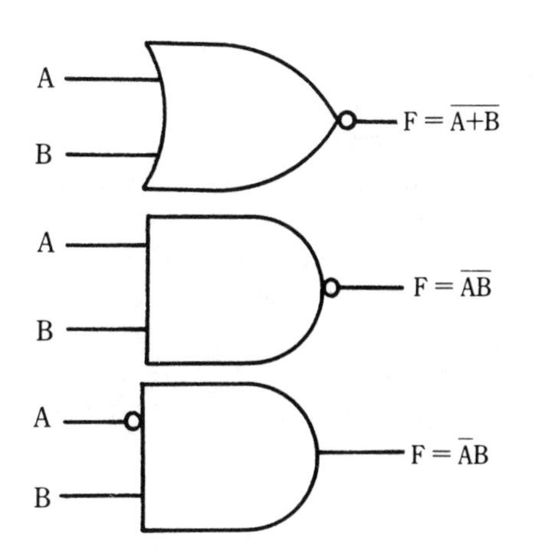

11-10 Inverter symbol used with logic symbols.

Combinational logic

These gates can be seen on logic drawings in many combinations. The combination for Q6 through Q10 is shown in Fig. 11-2. Because gate 2 is an OR gate, the output, G, should be C OR D. Hence, $C + D$ is the correct representation for G. The answer then to Q6 is d, None of these.

In Q7 it is given that A is a logic 1 and that B is a logic 1. The NAND gate should, therefore, have a logic 0 output.

Positive logic is used in Q8. A, B, and C are, therefore, "0" while D = "1." Because $G = C + D$ then G = "1" + "0" which is equal to "1."

$$F = \overline{\mathrm{AB}} = \overline{0 \bullet 0} = \overline{0} = 1$$
$$H = F + G = 1 + 1 = 1$$

The answer then to Q8 is logic "1" which = +5 volts.

By similar logic, the answer to Q9 is b, logic 0.

To figure out Q10 let's first take each gate individually.

$$F = \overline{AB}$$
$$G = C + D$$

and because $H = F + G$ by substitution then

$$H = \overline{AB} + (C + D)$$
$$= \overline{AB} + C + D$$

Working with equations such as these is Boolean algebra. There are many books available for further study in this field. See the bibliography at the end of this chapter.

Flip-flops

Switching circuits that are arranged to have feedback by cross-coupling collectors to bases are called multivibrators (Fig. 11-11). This type of feedback is regenerative in the active portion of the transistor's operating range. The networks in 1 and 2 of the circuit of Fig. 11-11 can cause the multivibrator to be astable, bistable, or monostable. If the coupling networks are primarily resistive the circuit will be bistable. If the networks are primarily capacitive the circuit will be astable (an oscillator). If one network is capacitive and the other resistive the circuit will be monostable (one stable state). All three are used in digital circuitry. All flip-flops are bistable multivibrators. Therefore, the answer to Q11 is c. Another name for the monostable multivibrator is the one shot. See Fig. 11-12. A special bistable circuit is the Schmitt trigger. This circuit changes state when the single input is above or below a particular level. It is often used to "square up" slow changing signals.

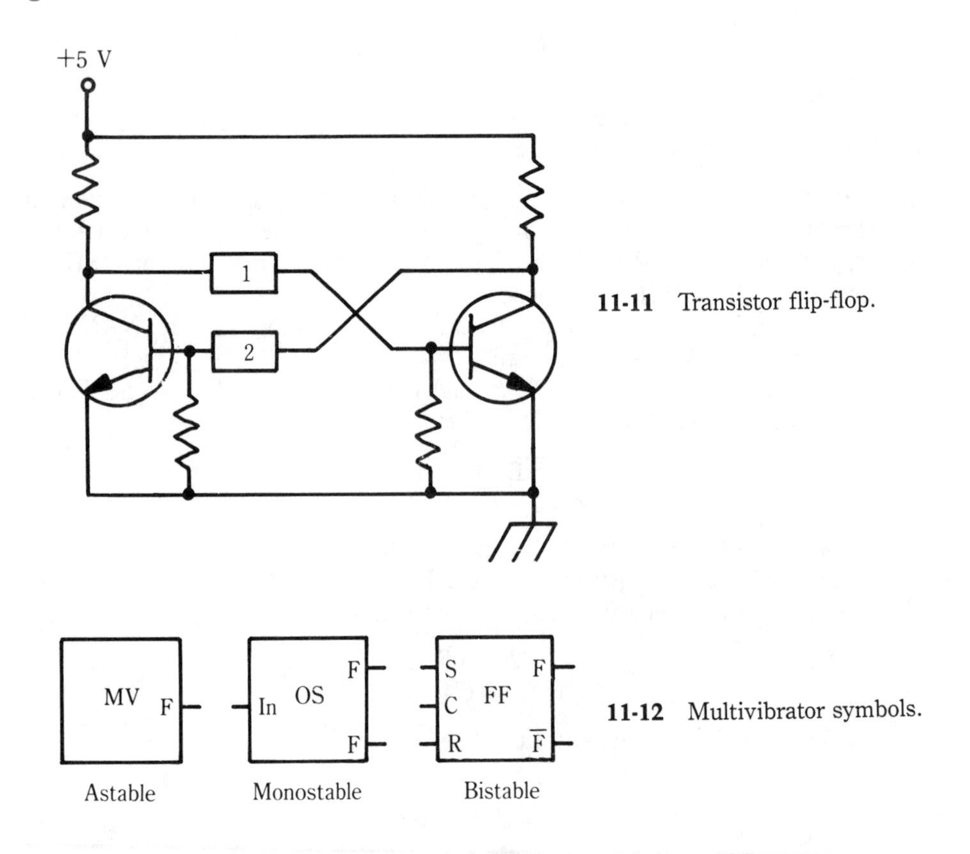

11-11 Transistor flip-flop.

11-12 Multivibrator symbols.

There are several types of flip-flops. Perhaps the simplest of these is the set-reset flip-flop without a clock input. It is shown symbolically in Fig. 11-13. Notice that it contains four signal lines (power connections are seldom shown). The inputs are normally shown on the left. They are *S* for the set input and *R* for the reset input. The outputs are often named for the particular function of the flip-flop. *F* designates function. For example, if the flip-flop were the fourth flip-flop in a data register, the outputs might be labeled *D*4 and $\overline{D4}$. The outputs are derived from the collectors of the transistors. If one transistor is ON, the other is OFF; hence,

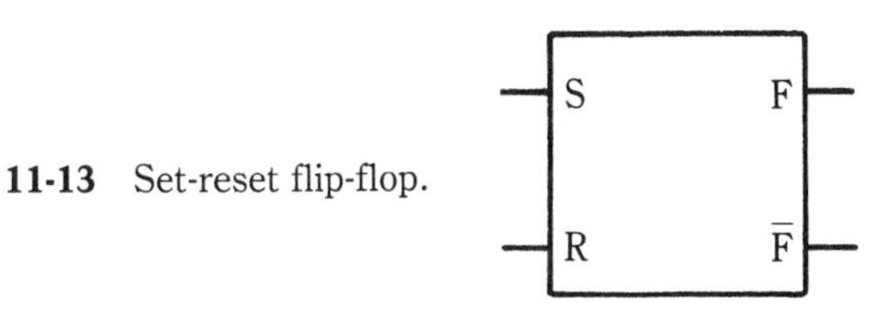

11-13 Set-reset flip-flop.

the outputs are always complementary. That is, when $F=$ "1," $\overline{F}=$ "0." When $F=$ "0," $\overline{F}=$ "1." A flip-flop is said to be *set* when the F output is a logic 1. The flip-flop is said to be *reset* when the F output is a logic 0. See Fig. 11-14.

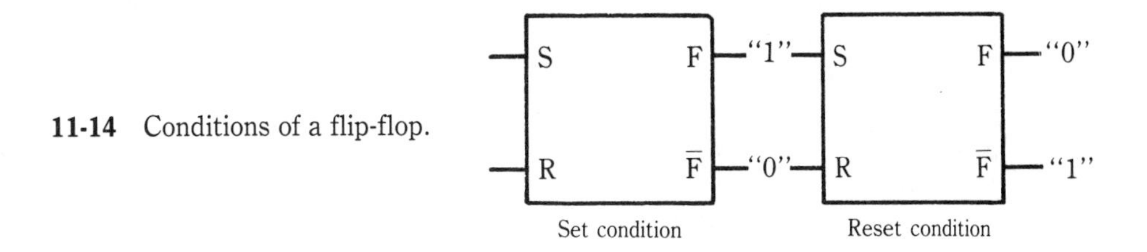

11-14 Conditions of a flip-flop.

A set-reset flip-flop will be *set* after the application of certain input conditions, namely $S=$ "1," and $R=$ "0." It will be *reset* after application of the conditions that $S=$ "0" and $R=$ "1." If S and R are both logic 0, the flip-flop will stay in whatever condition it was in. The condition of a logic 1 on both inputs is forbidden. These conditions are summarized in technical manuals with a truth table. The table needs to account for all possible combinations of input and the *present* state of the flip-flop to describe how the flip-flop will react to the inputs. Let's use F to represent the *present* state of the flip-flop and F' to represent what the flip-flop will go to after the application of the input conditions. This together with applying the behavior of the set-reset flip-flop will give the following truth table (Table 11-1).

Table 11-1. Truth table for S-R flip-flop.

Notes	R	S	F	F'
No change	0	0	0	0
No change	0	0	1	1
Output sets	0	1	0	1
Output stays set	0	1	1	1
Output stays reset	1	0	0	0
Output resets	1	0	1	0
Not allowed	1	1	0	ϕ*
Not allowed	1	1	1	ϕ*

* ϕ is a symbol that means this condition is forbidden. It is formed by the combination of the 0 and 1 overlayed. If 1's are applied the outputs are unpredictable.

Perhaps an easier way to show the flip-flop action is to use a slightly different truth table. In this truth table the F output is shown in the "future" by labeling the column F_{t+1}. This means that the column will contain the state of the F output at time $t+1$; "t" is implied as the time the inputs are applied.

R	**S**	F_{t+1}
0	0	F_t
0	1	1
1	0	0
1	1	0

If the flip-flop has a clock input the change of the output is inhibited until the clock occurs. Usually the edge of the clock pulse causes the output transition and either edge can be used. Some flip-flops trigger on the positive-going edge of the clock input and some trigger on the negative-going edge. The tables are the same. In the first table, F goes to F' when the trigger occurs. In the second table "t" can be thought of as the trigger time. For example, when R is 0 and S is 0 on the first line above, the output will stay at whatever it was at time "t"; i.e., F_t. If it was "1" at time t it will stay 1 at $t+1$.

In the circuit of Q12 the S input is tied to +5 volts and R input to 0 volts. If +5 volts = logic 1 then 0 volts is a logic 0. A "1" on the S input and a "0" on the R input will set the flip-flop, so F output will go to +5 volts, logic 1. (Note that if negative logic is used, the output still goes to +5 volts, but it would be called a logic 0.)

Another common type of flip-flop is the J-K flip-flop. It is symbolized in Fig. 11-15. J can be thought of as the set input and K as the reset input. The letter Q is not unique; it is often replaced with a symbol significant of the flip-flops function. The J-K flip-flop responds exactly the same as an R-S flip-flop except that J and K inputs can be logic 1 at the same time. When they are, the flip-flop will complement. If Q was a logic 1 (set), it will go to logic 0 (reset). Each table form is shown here.

K	**J**	**Q**	**Q′**
0	0	0	0
0	0	1	1
0	1	0	1
0	1	1	1
1	0	0	0
1	0	1	0
1	1	0	1
1	1	1	0

K	**J**	Q_{t+1}
0	0	Q_t
0	1	1
1	0	0
1	1	$\overline{Q_t}$

Both tables are repeating the same rules that applied to the S-R flip-flop except when J and K are both logic 1. That is: the flip-flop remains in whatever state it was in if J and K are "0," it sets if J = "1" and K = "0," it resets if J = "0" and K = "1" and it *toggles* if J and K are both "1." If the flip-flop has a clock input the output will be affected only at the appropriate trigger time (either leading or trailing edge can be used for trigger).

Many J-K flip-flops are internally arranged such that the omission of connections to the J-K inputs will be the same as tying J and K to logic "1." Hence, without the J-K inputs the flip-flop will toggle when a trigger is applied. (At successive trigger pulses the output will alternate between "1" and "0.")

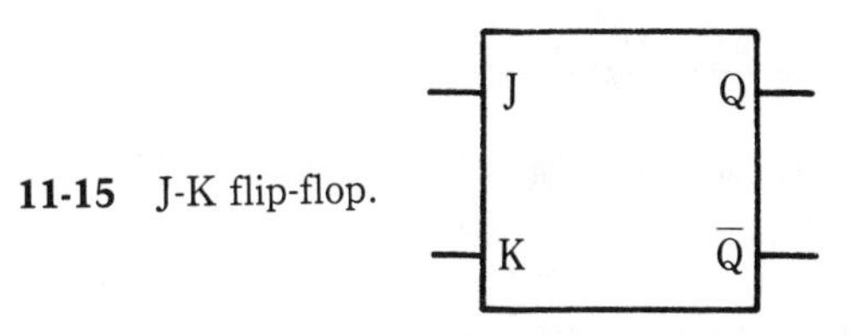

11-15 J-K flip-flop.

A toggle flip-flop behaves like a J-K flip-flop with only a clock input. The toggle flip-flop symbol is shown in Fig. 11-16 with its truth table. The answer to Q13 is true.

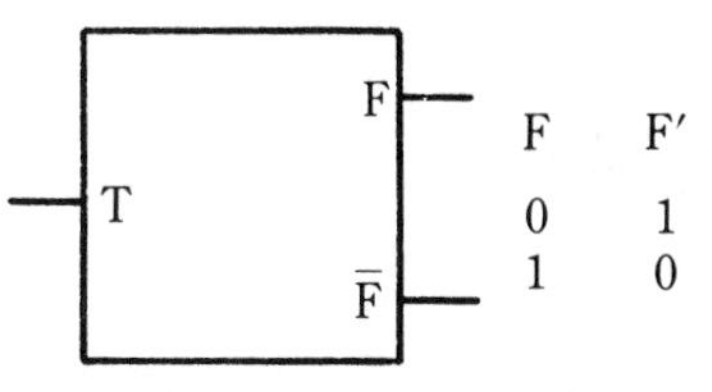

11-16 Toggle flip-flop and truth table.

Another flip-flop is called the data flip-flop. It usually has a clock and it has only one logic level input as in Fig. 11-17. It behaves in the following manner. A logic 1 on the input will become "stored" in the flip-flop when it is clocked. That is the output will go to a logic 1 after the trigger. A logic 0 will likewise be "stored." The truth tables for a data flip-flop are shown here.

D	F	F'
0	0	0
0	1	0
1	0	1
1	1	1

D	F_{t+1}
0	0
1	1

Data flip-flops, as the name implies, are used to store data. They are used in registers or memory. Question 14 is, therefore, answer d. The data flip-flop is often incorporated in large scale integrated chips (LSI chips). When so made, the result is a random-access memory or RAM.

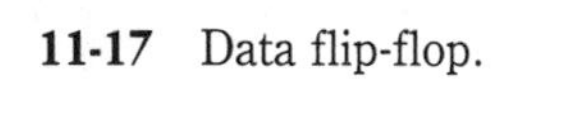
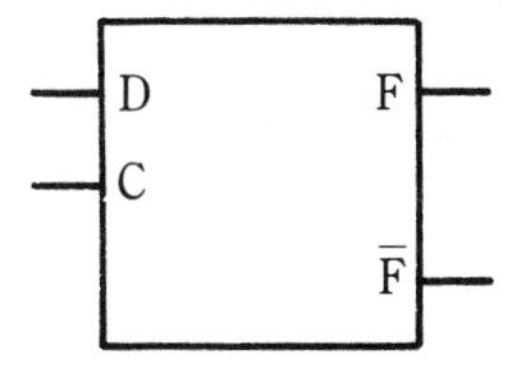

11-17 Data flip-flop.

Additional reading

Horn, Delton T.: *Using Integrated Circuit Logic Devices*, TAB Books, 1984.
Larson, Boyd: *Transistor Fundamentals and Servicing*, Prentice-Hall, 1974.
Maloney, Timothy J.: *Industrial Solid-State Electronics*, Prentice-Hall, 1979.

Namgostar, M.: *Digital Equipment Troubleshooting*, Reston Publishing, Inc., 1977.

Paynter, Robert T.: *Microcomputer Operation, Troubleshooting and Repair*, Prentice-Hall, 1986.

Slater, Michael and Barry Benson: *Practical Microprocessors, Hardware, Software, and Troubleshooting*, Hewlett-Packard Co., 1979.

12 Safety precautions

THIS CHAPTER COVERS THE TYPES OF SAFETY QUESTIONS THAT WILL BE asked on the exam.

Quiz

Q1. A shorted balun on a TV set or other communications gear, coupled with a "hot" chassis, is safe if the equipment uses a polarized line cord:

a. True
b. False

Q2. A spark obtained when connecting antenna leads or cables to a receiver antenna post can be caused by:

a. static electricity collected on the antenna elements.
b. leakage between windings in the receiver power transformer.
c. receiver on/off switch leads being hooked up backwards.
d. antenna leads being reversed.

Q3. Which of the following will be *least* likely to present a shock hazard to a repair technician?

a. radio transmitter final output-stage tube-plate connections.
b. audio amplifier regulated power supply transformer wires.
c. tube-type TV receiver p-c board connections.
d. mast-mounted RF signal amplifier circuit.

Q4. CRT high-voltage power-supply rectifier tubes can be safely removed, without shock hazard, providing the cathode circuit—CRT pigtail wire circuit—is not brought into close contact:

a. True
b. False

Q5. The CRT high-voltage circuit should be discharged by:

a. shorting to chassis.
b. shorting to the cabinet.
c. using a high-voltage probe.
d. shorting high voltage to collector heatsinks.

Q6. To prevent further damage when troubleshooting a receiver that appears to have an internal short:

a. use a fuse less than 100 percent larger than the recommended size.
b. use a dropping resistor in series with the ac line or battery voltage.
c. substitute a resettable circuit breaker for the line fuse.
d. use an isolation transformer.

Q7. Equipment used to monitor medical patients can develop leakage current. To reduce the likelihood of leakage between the patient and the equipment:

a. use equipment with a 3-prong to 2-prong ac adapter.
b. ground the patient.
c. use only AHA equipment.
d. use equipment (containing an isolation transformer) that had a leakage test prior to use.

Q8. It ordinarily takes 100 – 500 mA of 60 Hz ac current arm-to-arm to induce ventricular fibrillation. Lower levels of current also can induce fibrillation if:

a. the person is isolated.
b. higher frequencies are induced.
c. the person is hypertensive.
d. medical equipment (needles, contact probes, etc.) is connected to the person causing a lowering of his/her skin resistance.

Q9. Equipment used in dangerous locations (such as damp basements, construction sites, outside the home, etc.) can *best* be made safer by:

a. operating with one hand only.
b. checking resistance between line cord and case.
c. cease using if case becomes "hot."
d. using ground fault ac circuits for power.

Q10. Which of the following tests can be performed without damage to the component?

a. shorting collector to base on a transistor.
b. shorting collector to emitter.
c. shorting base to emitter.
d. shorting emitter to ground.

Q11. CMOS chips, field-effect transistors and other sensitive electronic components can be damaged easily by:

a. operating for long periods of time.
b. using near high-voltage supplies.
c. long shelf life.
d. any type of static discharge.

Equipment shock hazards

One purpose of the polarized ac plug is to assure that the electrical equipment chassis of a "hot chassis" is connected to the low side of the ac power source. By so doing the shock hazard is lessened. An unsafe condition occurs if the balun of a TV set is shorted. The balun, made up of a couple of capacitors and a few turns of wire around a ferrite sleeve, is used as a transformer. Baluns are susceptible to lightning damage. Should the receiver chassis be connected to the hot side of the line, and should that chassis be shorted to the antenna system through the balun, a shock hazard exists. See Fig. 12-1.

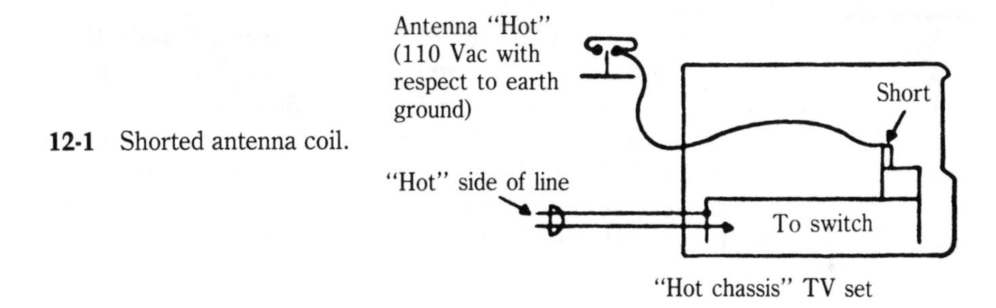

12-1 Shorted antenna coil.

While use of polarized line cords lessens the chance of shock hazards of this kind, some ac power outlets have been known to have been wired backwards! Therefore, Q1 is a false statement. Other equipment (such as an add-on do-it-yourself phonograph) that can be connected to the primary unit might have ac power shorts also. The best rule then is to never completely trust any equipment you are working on. *Always* find the cause of sparks when the antenna cable or leads are connected or when any other equipment is connected.

In addition to the previous discussion, power transformers (especially after years of use in a hot environment) might have insulation breakdowns that cause the primary ac input windings to be connected to one or more of the secondary windings. The answer to Q2 is b. Depending on the resistance of the short and the location, the receiver can become "hot." When this occurs, a 50-Vac to 120-Vac level on the chassis is not unusual. This creates a severe shock hazard and must be repaired, or the equipment taken out of use. Measuring the ac levels between power source, ground, and chassis is recommended practice for all electronics technicians.

Transmitter output tubes are high-power circuits using high voltage. Observe extra precautions here. Except in small transistor amplifiers using low dc levels, amplifier power supplies can be designed to produce 100 volts to 200 volts and more. Present day regulator circuits with their high-current capability can be hazardous. Tube type TV and radio circuits often use 100 volts to 300 volts supplies and can use 110 volts ac to power the filament circuits. Few areas of these sets do not pose a serious shock hazard. The rule is to use only one hand in any of these sets with the other hand completely isolated from any part of it.

The mast-mounted RF signal amplifiers used in TV and radio use ac power from the ac-line source, but this is dropped through a transformer to about 20 Vac before being supplied through the coax or antenna down-drop to the mast-mounted transistorized signal amplifier. This small source is not much of a hazard. The *best* answer to Q3 is d, but should the source have a fault, such as a shorted step-down transformer, then this too becomes a shock hazard.

Although technicians usually regard 15 volts to 25 volts power sources as little or no shock hazard, a note here is in order. Some of the most serious accidents involving electrical shorts have come in low-level circuits. Automobile radios are an area where technicians have been known to get their rings or watch bands shorted between 6 volts or 12 volts and the frame. With little resistance between the frame, the battery, and the solid piece of gold, silver, or other conducting metal, the heat generated can cause a severe burn in seconds.

Most of the time the statement in Q4 is true. The cathode circuit and CRT pigtail are usually well insulated to eliminate arcing and secondarily to protect the technician. However, a percentage of high-voltage-rectifier vacuum tubes and some solid-state diode types will develop shorted elements. Taking off the anode cap and coming in contact with the anode is then the same as touching the cathode CRT pigtail circuit. See Fig. 12-2. It can be a shock hazard. Grounding these circuits before servicing is the surest method of avoiding shock. And, of course, the set should be turned off.

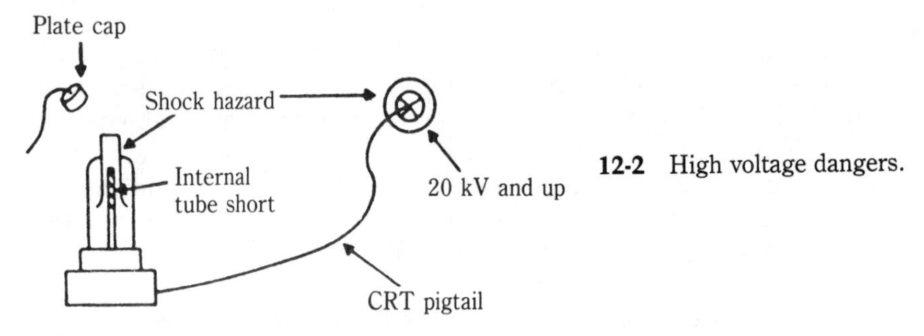

12-2 High voltage dangers.

All video screens use high voltage on the second anode. This might be as high as 40 kVdc. Television sets and computer monitor screens are becoming even more numerous and subject to service. The potential on the high-voltage tube (such as 3A3) or solid-state tripler or quadrupler can be harmful. Even if the actual current is not harmful in itself, the physical reaction by the technician to a shock can cause harm to himself, or to other equipment. Picture tube circuitry is such that the high voltage can be partially maintained for some time before it can bleed off. In Q5 a high-voltage probe can be used to discharge the circuit prior to servicing, but the proper method is to short the pigtail lead to chassis ground. It should *not* be shorted to the cabinet because the cabinet might *not* be directly connected to the chassis and the shock hazard will still exist. It should not be shorted to any heat-sinks used to mount transistors as they may be isolated from the chassis and the transistor might be destroyed by the overvoltage.

Troubleshooting shorted power problems

The technician must learn how to service equipment that has an internal short, one that perhaps blows fuses. Some technicians use a circuit breaker and clip leads with small diameter wire, as in answer c of Q6. The circuit breaker protects against any large current and the small wire also acts as a resistance and fuse. This protects the set and can quickly show the location of the short. While this practice might be acceptable where the equipment being serviced is familiar, a better method for all kinds of equipment is to insert a large-wattage dropping resistor in series with one leg of the ac power line. The more severe the short, the greater the drop across the resistor, limiting the current and thus the damage that might be caused to the equipment. A 50-ohm 10-watt resistor will work on sets that are designed to operate up to 1 amp of 110-Vac power. Most technicians use a variac to troubleshoot shorts, gradually increasing the ac power input while monitoring the unit for the short and for the input current to keep it reasonable.

Medical electronics safety

Using an isolation transformer between equipment and patients eliminates some leakage possibilities. Obviously, ground fault ac circuits and remote sensors isolated with optical couplers are good protection devices. Many technicians use isolation transformers to power their shop workbenches and equipment in a like manner to eliminate such hazardous potentials. Never defeat the 3-prong plug with an adaptor in medical environments. The answer to Q7 is d.

In medical monitoring equipment, probes can be placed on the skin or actually in the body. The resistance of the body at contact points is reduced. Therefore much smaller levels of current are needed to cause severe patient harm. The answer to Q8 is d.

Safety environment

Question 9 is about equipment used in dangerous locations. Ground fault interrupters open the line-voltage supply with the slightest amount of leakage current caused by: defective tools and equipment, stripped insulation on ac service lines, or other causes. Technicians who are routinely located in shops or field locations with wooden floors and benches, rugs, and plastic floor coverings become complacent regarding shock hazards. Other environments (out-of-doors and in locations where wetness occurs) can place the technicians and others in lethal positions. Extra care should always be used in these areas.

Transistors are difficult to service. Question 10 addresses this difficulty. A momentary short of the test lead between collector and emitter or between collector and base will destroy the transistor. A simple trouble can be made into a compound one if parts are ruined by carelessness. The only sure method of turning transistors on and off safely is by carefully shorting the base to the emitter. This removes the forward bias and turns off the transistor. Caution should be used in direct-coupled stages with this method.

Static is a constant problem with sensitive CMOS chips. Merely walking a few steps on a nylon carpet can build up too many volts of static electricity on a technician. Should he/she then touch the chip or transistor leads and cause the static to discharge through the chip in the wrong way, the part will be destroyed as the static blows a hole in the very thin insulation between component sections. The answer to Q11 is therefore d.

National Electrical Safety Code

Technicians also need to keep track of safe installation practices when installing equipment in customers homes. Safety procedures, contained in equipment manuals, should be followed on all installations. In addition, The National Electrical Safety Code should be followed as well as all local codes governing the installation of electrical equipment and appliances. A portion of the code covers antenna installation, see Fig. 12-3.

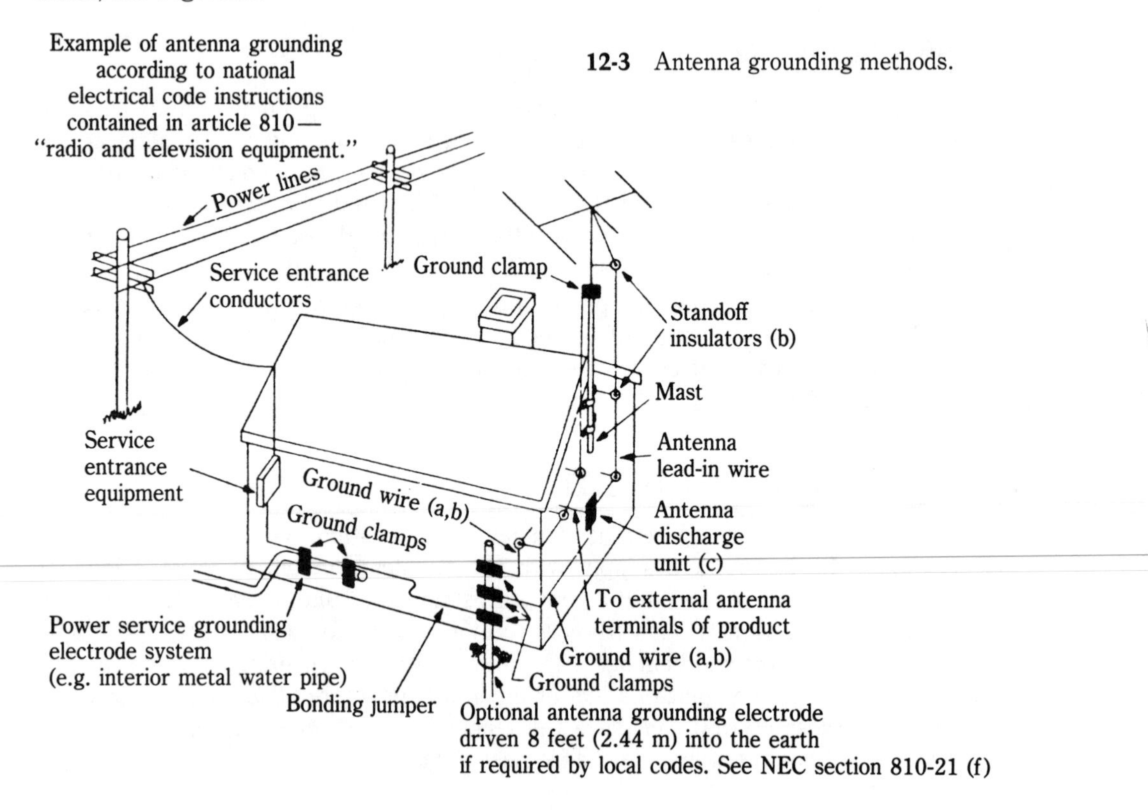

12-3 Antenna grounding methods.

Additional reading

Lacy, Edward A.: *The Handbook of Electronic Safety Procedures*, TAB Books, 1983.

Lenk, John D.: *Handbook of Practical Microcomputer Troubleshooting*, Reston Publication Co., Inc., 1979.

Summers, Wilford J.: ed. *The National Electrical Code Handbook*, National Fire Protection Association, 1978.

13 The consumer electronics option

Quiz

Q1. A carbon microphone is similar in operation to a/an

a. phonograph cartridge.
b. electromagnetic speaker.
c. variable capacitor.
d. variable resistor.

Q2. Loudspeaker impedance is:

a. the resistance measured on the 1-ohm scale of a VTVM.
b. important for matching purposes.
c. the impedance in ohms measured at 10 kilohertz.
d. the average impedance over the 16 hertz – 20 kilohertz audio range.

Q3. Phonograph cartridges can be expected to produce what level of output?

a. 1 volt for ceramic; 5 – 10 millivolts for magnetic.
b. 100 millivolts for ceramic; 50 millivolts for magnetic.
c. 5 volts for ceramic; 100 millivolts for magnetic.
d. both types produce equal levels of output.

Q4. RF signals are often measured in microvolts. With an input to a stage of 300 microvolts, and a voltage gain of 6, the output voltage of the stage would be:

a. 900 microvolts.
b. 9 millivolts
c. 1.8 volts
d. 1.8 millivolts

Q5. Match the following commonly used IF (intermediate frequencies) to their product or circuit:

a. 455 kHz	1. TV sound frequency
b. 10.7 MHz	2. TV video IF
c. 45.75 MHz	3. FM radio
d. 4.5 MHz	4. AM radio
e. 19 kHz	5. FM stereo pilot
f. 67 kHz	6. SCA (subsidiary communications authorization)

Q6. Match the waveforms in Fig. 13-1 to the descriptions below:

a. TV signal
b. AM radio signal
c. FM radio signal
d. audio waveform

(1)

(2)

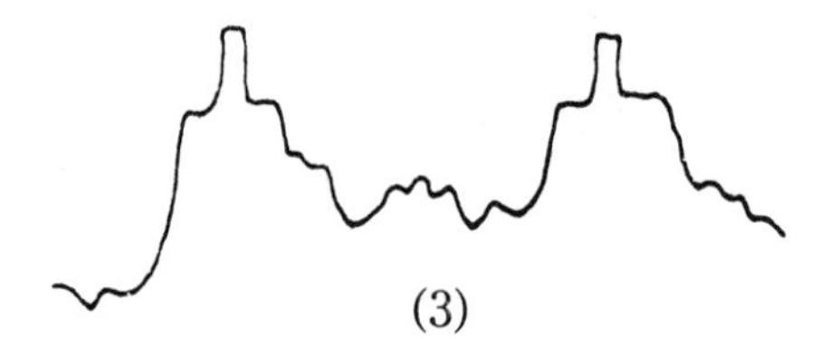

(4)

13-1 Waveforms for matching in Q6.

Q7. Match the following transmission bands to the frequencies listed:

a. 1 MHz	1. FM band
b. 4 GHz	2. TV band
c. 54 MHz	3. AM broadcast band
d. 96 MHz	4. satellite receiver band

Q8. When confronted with an AM/FM cassette player that performs OK on tape but has no AM or FM reception, a good first step in finding the trouble is to check the:

a. power supply voltages
b. audio output and speaker circuit
c. external antenna connections
d. tuner and IF sections

Q9. The doubler in Fig. 13-2:

a. is the standard squelch circuit.
b. doubles the audio voltage.
c. doubles the agc voltage.
d. doubles 19 kHz to 38 kHz.

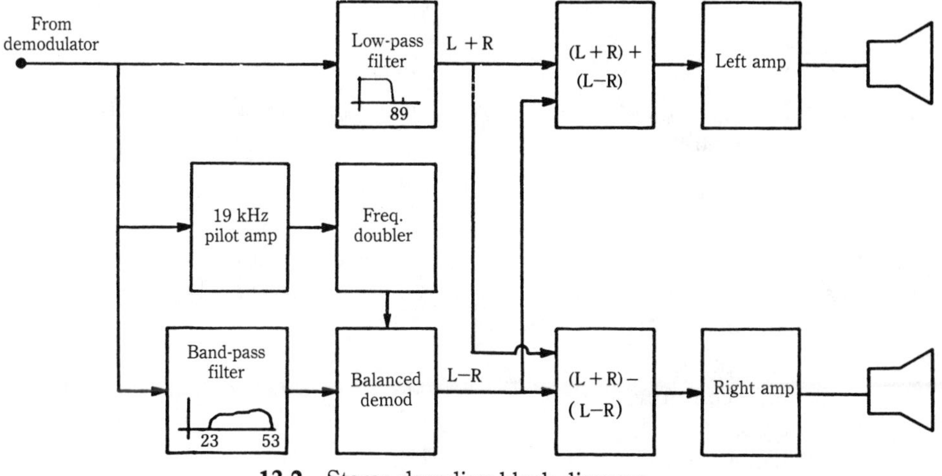

13-2 Stereo decoding block diagram.

Q10. A degaussing coil is used to demagnetize:

a. the permanent magnets in the yoke of a TV set.
b. the shadowmask inside the picture tube.
c. the conductive coating inside the bell of the kinescope.
d. the metal portions of the TV cabinet.

Q11. A balun as used with TV sets is:

a. a matching transformer.
b. a VHF/UHF bandsplitter.
c. an ac power isolator and cable TV conversion switch.
d. ac power connection interface plug.

Q12. 41.75 MHz video IF, and 45.75 audio IF are common TV intermediate frequencies

a. True
b. False

Q13. The video drive signal to the 25″ CRT control grids, or cathodes, should be about what level?

a. 1000 microvolts
b. 1 to 5 volts
c. 100 volts
d. 1000 volts

Q14. Twin lead TV signal distribution cable should never be used when connected to 35 channel cable TV systems.

a. True
b. False

Q15. In cable-TV systems, channel 14 (A) is immediately above channel 7 in frequency.

a. True
b. False

Q16. The video peaking and the "picture" control on many TV sets do the same thing functionally.

a. True
b. False

Q17. The blank TV screen raster is white except for the lower-right corner, an area about the size of a baseball on the 25-inch screen. This area is pink. What adjustment might correct the condition?

a. dynamic purity
b. dynamic convergence
c. static purity
d. static convergence

Q18. U1 has a gain of 10. What can be stated further about the circuit in Fig. 13-3?

a. U2 has a gain of 5.
b. The entire circuit has a gain of 15.
c. The speaker cannot work without a ground.
d. The feedback resistors on the 741s should be connected from B+ to pin 2, rather than from the outputs.

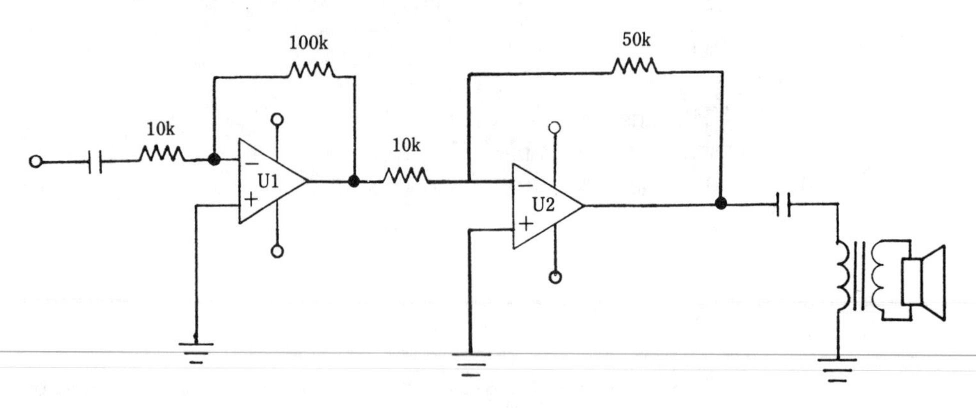

13-3 Drawing for Q18.

Q19. Identify the components in Fig. 13-4.

a. A erase head; B video head; C capstan pressure roller

b. A video head; B audio head; C erase head
c. A audio head; B video head; C erase head

Q20. I, Q, and Y signals are used in Video recording. Which is the luminance signal?

a. I
b. Q
c. Y

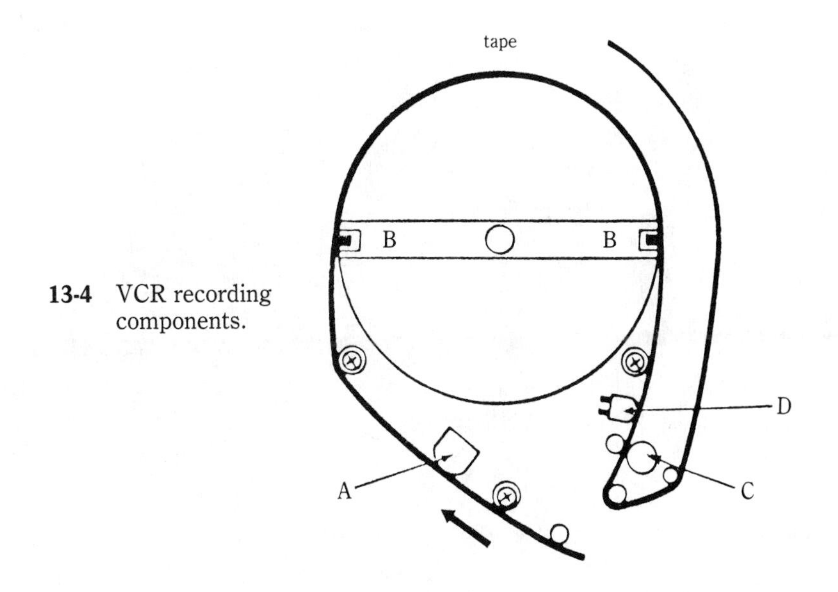

13-4 VCR recording components.

Quiz review

Question 1 requires you to understand how a carbon microphone works. In this case, the carbon mike acts like a variable resistor. The compacted carbon granules are compressed as sound vibrations put pressure on the attached diaphragm. The variations in resistance are transferred to an amplifier.

Electromagnetic speakers are not popular today. They operate in the same fashion as the common permanent-magnet speaker. They use a speaker coil operating in the circular gap of a strong magnet. The audio power concentrated in the voice coil magnetic field causes a motor action. This action drives the speaker diaphragm at the audio rate. An electromagnetic speaker uses a large coil, rather than a permanent magnet, to produce the magnetic field for the speaker voice coil to operate in. Early radios sometimes used the field coil of the speaker as part of the power supply filter circuit, thus accomplishing two tasks with the single electromagnetic coil.

Phonograph cartridges, if magnetic, operate like speakers in reverse. The phonograph track vibrations are picked up by the jewel needle, which is attached to a small magnet moving within a coil. While the output voltage of a magnetic cartridge is very low compared with a crystal or ceramic phono cartridge, the fidelity is normally better. Magnetic phono cartridges produce less than 10 millivolts of signal, while crystal and ceramic cartridges output a volt or more—thus requiring no preamplification as does the magnetic.

Capacitance microphones were a type that used a ribbon of thin metal which vibrated slightly with sound waves and had an improved frequency response compared with early carbon mikes (Fig. 13-5).

13-5 Dynamic (voice coil in permanent magnet field), microphone (left) and headphone speaker (right).

Loudspeaker impedance is not the resistance of the voice coil. It is the combination resistance and inductance at audio frequencies. Technically it is defined as the lowest impedance amount that can be measured at frequencies above the speaker's resonant frequency.

In Q3 we are dealing with phono cartridges, which with CDs and DAT (digital audio tape) are not so important as in the past. However, phono cartridges will be with us for a decade of so, at least. Ceramic cartridges are high-output types as stated earlier. Answer a is correct for Q3.

Question 4 asks you to express gain in an amplifier. We aren't dealing with dBs here, only straight numbers. A gain of 2 means the stage doubles the input voltage. The output is twice as large as the input, or 2 times as large. A voltage gain of 6 is 6 times the input. With 300 microvolts in, the output is 1800 microvolts, or 1.8 millivolts as in answer d.

Question 5 asks you to match common frequencies every tech should know.

The standard AM radio IF is 455 kHz. The standard FM radio IF frequency is 10.7 MHz. The common video IF carrier frequency in TV is 45.75 MHz. Also in TV, rather than extracting, or processing the 41.25 audio FM carrier which accompanies the video in a TV IF circuit, a much lower frequency 4.5 MHz (the difference between the 45.75 and 41.25 audio IF) audio difference beat frequency is derived at the detector and processed by the sound IF circuitry. Therefore the audio subcarrier in TV is 4.5 MHz.

Stereo FM radio broadcasts, rather than using two distinct carrier frequencies, use only one. An FM stereo pilot frequency is used to create a second carrier for the second channel. The 38-kHz pilot isn't broadcast. It is eliminated prior to broadcast; however, the sidebands, or useful information, is broadcast within the FM signal. In the receiver a 19-kHz oscillator is used to double 19 kHz to 38 kHz, thus providing the carrier in the receiver where the left and right signals are extracted.

SCA is used to allow subcarriers on ordinary FM broadcasts. These can be used to provide auxiliary audio information channels, such as those used for background music services. The key frequency used is 67 kHz.

The troubleshooting question, Q8, asks what is most practical: To randomly check this and that, or to logically realize if the cassette player has audio and drive functions, the power supply is most likely OK. Because the radio function is inoperative, something between the antenna and the detector is bad. c and d are good answers. Antenna connections are a quick and easy thing to do first. If no trouble is found there, inspect the tuner and IF sections.

The answer for Q9 is d.

Question 10 asks if you know exactly what a degaussing coil does. Each modern TV set has a degaussing coil permanently attached around the faceplate of the CRT (cathode ray tube). Each time the TV is turned on cold, the circuit is energized and the shadowmask, directly behind the phosphor dots on the color screen, is demagnetized. Occasionally, due to a large-magnet speaker placed adjacent to the TV, the CRT mounting frame, cabinet, and shadowmask are strongly magnetized and a degaussing coil as carried by most TV techs is needed to purify the screen discolorization. This is the same symptom as is caused by a slightly bent shadowmask. The bent shadowmask problem is unlikely to be improved on. b is the best answer to Q10.

For Q11 you must know that BALUN means bal-un, or balanced/unbalanced. A balun is a transformer that matches 300-ohm twin lead to 75-ohm coax, or some other differing impedance components.

In Q12 the video and audio TV IF carrier frequencies are reversed. The answer is false.

For Q13, if you have no idea what the signals driving the CRT should measure, you could see a proper "looking" waveform (on the CRT cathodes or grids), but the amplitude might be too small. Remember that big CRT tube requires a big signal, 100 to 150 volts.

Twin lead, unshielded transmission cable was commonly used to wire extra rooms in homes. However, cable TV systems invariably use Midband channels A,

B, and C, which are also used for aircraft communications. (Superband channels K and M also use aircraft frequencies.) FCC rules required cable companies to not distribute their signals on systems that might emit these frequencies, thus interfering with aircraft communications. A is correct for Q14.

Just about all techs work with RF signals in some way. Even if you merely watch TV, you should know what frequencies are being used. This question is false because channel 7 in TV is right above the midband A-I channels used by cable. Superband is above channel 13.

Video peaking (Q16) is called sharpness on many TVs, but more and more is being named PICTURE control. It is a control that reduces the high-frequency response of the video amplifier, prior to the signal reaching the CRT. The effect is to eliminate graininess and too-sharp edges on objects in the video.

Question 17 carried on the subject of Q10, asking if you recognize impurity on a TV screen. The not-pure TV screen can be recognized by a pink or green or blue smudge, instead of a pure white screen, when the video signal is removed. An impure screen is caused by the respective CRT electron gun beams being influenced to move from the shadowmask aperture holes they are intended to travel through, and instead, wandering to an adjacent wrong hole. Therefore, if the beam from the blue gun falls on the red phosphors, the red is activated instead of the intended blue, causing a pink or red smudge. When this condition occurs, many aperture holes are affected in perhaps a baseball sized section of the TV screen.

You must understand how the gain of an op-amp is designed for Q18. U1 has a 100-k feedback resistor and a 10-k input resistor. This results in a gain of 10. U2 has a 50-k feedback and a 10-k input, resulting in a gain of 5. The entire circuit then has a gain of 50. a is correct.

In Q19 a is the erase head, b the video head, and c the capstan pressure roller.

In Q20, the Y signal is the luminance or brightness portion of the TV signal, with I and Q containing the color information: hue and intensity.

Other consumer product knowledge

As a TV or consumer product technician, you also should become familiar with antenna and satellite equipment as contained in this book and others (see chapters 17 and 20). If you are not familiar with the present-day power supplies used in TVs, make it a point to find out how they work. Protection shut-down circuits in TV are difficult to service, even when you know exactly how they are supposed to function.

TV tuning methods, remote control circuitry, and graphics circuits are commonplace now. If you don't know how they work, you will be attempting to do a most difficult job when troubleshooting. Because most modern radios, TVs, and VCRs use digital tuning, you might as well determine to understand just what happens when you press a button or turn a knob. Even remote hand units require an understanding of the operation and technology. How do you improve the response of the key pad? What causes batteries to fail in one or two days? How can you tell if

13-6 20 dB TV signal amplifier increases signal level on both VHF and UHF. Has 12 dB gain control.

13-7 Amplifier plugs directly into ac outlet.

the unit is transmitting? Can you troubleshoot them with an oscilloscope? The Consumer Option of the CET exam doesn't contain in-depth questions on every topic, but the modern exam will change with technology and your job is to keep up with it. See Figs. 13-6 and 13-7.

14 Computer basics

Quiz

Q1. A microprocessor-based system is:

a. bus oriented
b. versatile
c. controlled by software
d. All of these.

Q2. A byte is usually understood to be:

a. 4 binary bits
b. 8 binary bits
c. 32 binary bits
d. All of these.

Q3. Three-state drivers are often used to drive bus lines.

a. True
b. False

Q4. In microcomputers RAM is used for:

a. permanent storage
b. address decoding
c. temporary storage
d. None of these.

Q5. Assembly language is:

a. machine language
b. BASIC
c. mnemonic groupings
d. None of these.

Q6. A logic probe is used to probe points in the logic circuit of Fig. 14-1. The probe indicated a "1" at B, pulses at A, but logic 0 at F. This means that:

a. AND gate 1 could be bad.
b. AND gate 1 or OR gate 3 could be bad.
c. OR gate 2 is good.
d. None of these.

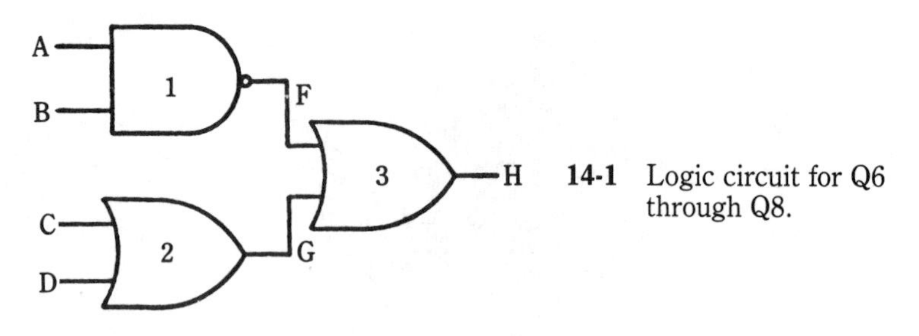

14-1 Logic circuit for Q6 through Q8.

Q7. In probing points C and D of Fig. 14-1 with a logic probe, both points indicated that pulses were present. G then should:

a. blink or be logic 0.
b. be logic 1 or blink.
c. be logic 0.
d. be logic 1.

Q8. In probing the circuit of Fig. 14-1 with a logic probe it was found that the probe blinked at F, G, and H. This shows that:

a. OR gate 3 is okay.
b. OR gate 2 is okay.
c. AND gate 1 is okay.
d. at least one input gets through to H.

Q9. In circuit of Fig. 14-2 input B was tied to ground and then a pulser was applied to A. A logic probe at the output showed a logic 1. Assuming positive logic:

a. This is normal.
b. The transistor is probably bad.
c. The diode at A is probably bad.
d. None of these.

Q10. A logic analyzer indicates:

a. the logic state of many points.
b. rise and fall times.
c. hex codes.
d. a and c.

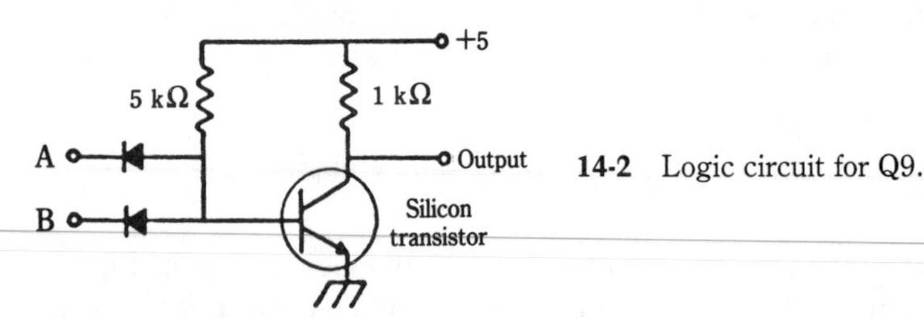

14-2 Logic circuit for Q9.

What is a microprocessor?

The microprocessor is an integrated circuit. It is a chip that contains all the computational and control circuitry for a small computer. It outputs and inputs information on pins that can be "bussed." Sequences of operations are controlled by "software." The answer to Q1 is d.

Fundamental principles

Microprocessors operate in a series of steps, moving along from step to step. At each step information on the bus pins stabilizes in some unique pattern of "1's" and "0's." Some pins are called data pins. These pins feed the data bus. These pins are sometimes inputs to the microprocessor and sometimes they are outputs to external circuitry. If there are eight of these data pins, the microprocessor is an 8-bit microprocessor. If there are four pins, it is a 4-bit microprocessor. There are 16- and 32-bit microprocessors also. The arrangement of data lines in groups of 4, 8, and 16 is related to the binary base of all computers. Because of the many times that it is necessary to refer to these groups, a system of "names" have evolved for these groupings. These names are tabulated here:

- A bit is one binary digit.
- A nibble is four bits.
- A byte is two nibbles or 8 bits.
- A word is 16 bits.
- A large word is 32 bits.
- A quadword is 64 bits.

The answer to Q2 is b.

Tri-state buffer

A *bus* is a line or connection that can be shared by various independent signal sources or signal destinations. Suppose two devices want to put a signal on a bus. You can see from the drawing in Fig. 14-3 that if A is high, Q1 would be ON holding the bus low. If B was low, Q2 would be OFF and (if Q1 wasn't there) the bus would go high. If A and B are going to both drive the bus, some accommodation

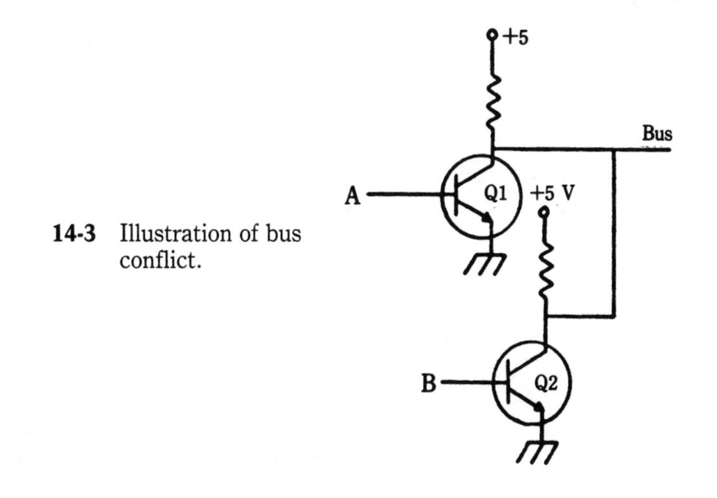

14-3 Illustration of bus conflict.

must be made to prevent this conflict. A three-state driver is used. That is, the output of the driver will be arranged so that it can be "1," or "0," or floating. One such circuit is shown in Fig. 14-4.

When the control input is low, D1 is forward biased holding Q3 base low. Q3 is OFF; therefore, Q4 is OFF. With control low the collector of Q1 is also low so Q2 must be OFF. Q5 is, therefore, also OFF. The output is "floating." With control high the output will follow the dictates of the logic input. If logic input is high, Q1 collector is high, turning Q2 ON. Q2 ON turns ON Q5 and turns OFF Q3, Q4. The output, therefore, goes low. If logic input goes low with control still high, the collector of Q1 goes low shutting Q2 OFF. With Q2 OFF the base of Q3 goes high turning Q3, Q4 ON. Q5 is OFF with Q2; hence, the output is forced high by Q3, Q4. In order to keep the bus from the paradox of trying to be high and low at the same time, no two three-state drivers can be enabled at the same time. Three-state drivers or logic elements with three-state outputs are, therefore, used to drive bus lines. The answer to Q3 is a.

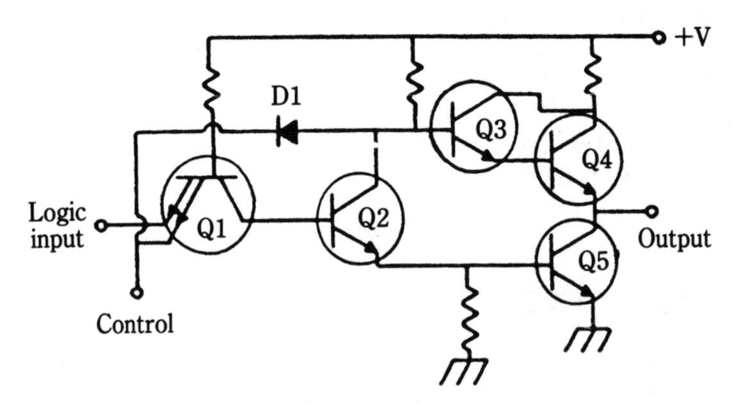

14-4 Tri-state driver.

Bus lines

Signal lines associated with a microprocessor are normally grouped into three categories: data bus lines, address bus lines, and control bus lines. These are shown diagrammatically in Fig. 14-5.

The 16 address lines (usually labeled A0 through A15) are controlled by a program counter in the microprocessor. If no instructions are brought into the microprocessor to change it, the program counter will count from zero to ($2^{16}-1$) or 65,535 and start over again. (The number of combinations possible in a 16-bit register is 2^{16}. Therefore, a counter can count from 0 to $2^{16}-1$.) The P.C.s (program counters) outputs feed the address bus lines, A0 to A15. Normally, however, information is brought into the microprocessor to alter this sequence. The P.C. can be made to skip counts or change to a previous count or go to any count desired. Addressing is, therefore, program controlled. This information is brought into the microprocessor by means of the databus lines, D_0 through D_7 (or D_0 through D_3 for a 4-bit microprocessor, or D_0 through D_{15} for a 16-bit microprocessor, etc.).

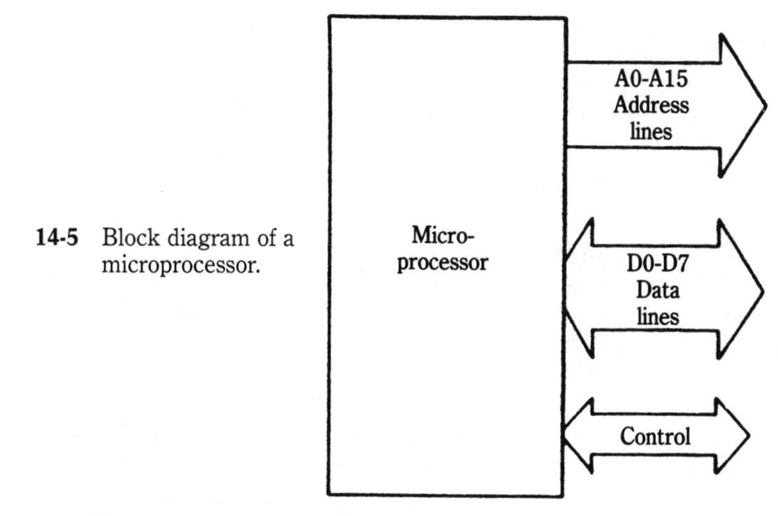

14-5 Block diagram of a microprocessor.

The microcomputer

The microprocessor by itself can do very little. It needs some external devices to make it into a usable microcomputer system, or microprocessor control system. The microcomputer needs a place to store the instructions to tell the microprocessor what to do. The microcomputer also needs a way to interact with people or other machines.

ROM

Read-Only Memory is used to store permanent program material. When a microprocessor is powered up, the program counter is normally reset to zero. The P.C. starts counting from there. The address lines, therefore, are all zero to start. A program is stored in ROM (Read-Only Memory), starting at address zero, that gets the microcomputer into whatever mode it needs to be in to do its work. This program is often called the resident monitor. If the computer has a keyboard, for example, one of the jobs the resident monitor needs to do is to "read" the keyboard to see if a key has been pressed. Other portions of the resident monitor, instruct the microprocessor on how to handle the information from the keyboard and still others arrange for the display.

RAM

The *Random-Access Memory*, RAM, is the read/write memory. That is, information can be stored there temporarily and later retrieved. Any information stored in RAM is lost when power is removed from the system. The answer to Q4 is c. Home microcomputers are often referred to by the amount of RAM available. For example, an 8 K machine, has 8192 bytes of Random-Access Memory available for temporary storage. When a particular address is written to, or read from, in RAM, 8 bits of data are involved, or one byte.

Language

The pattern of ones and zeros stored in the computer ROM or RAM is the only "knowledge" the computer has other than the pattern of wiring (the way things are hooked together). This sequence of bytes of data is a machine language program. When certain patterns of data are presented to the microprocessor on the data bus, the microprocessor "interprets" these patterns, decoding or operating on this machine language. Patterns of ones and zeros are very meaningful to the microprocessor then, but are pretty meaningless to people. To help people, higher level language or symbols are needed. Manufacturers of microprocessors list, in some fashion, the patterns of ones and zeros (the machine language) that are of particular importance to the microprocessor as instructions. The manufacturer also lists a mnemonic or name for each instruction and details of how the instruction will be interpreted by the microprocessor. The ordering of the instructions, along with data (necessary to some instructions) is called the machine language program. If the mnemonics are listed in the program order, the program is said to be listed in *assembly language*. Assembly language is not so easy to use either, so many machines have available a method of presenting and interpreting a higher level language, such as BASIC, FORTRAN, COBOL, or some other. Because BASIC is easy to use, it is often used in home computers. The answer to Q5 is c.

Troubleshooting microcomputers

There are many books written on the subject of troubleshooting microcomputers, and many books written on computers. The purpose of this section is to review only those troubleshooting subjects that every beginning technician should be on speaking terms with.

An oscilloscope is necessary in troubleshooting along with other meters. But one of the special tools pertaining to computers is the digital logic probe. This common tool is easy to use for switching circuits of all types. The logic probe is designed for a particular family of logic. TTL (Transistor-Transistor Logic) is widespread and, hence, logic probes are available for this family. CMOS (Complementary Metal-Oxide Semiconductor) families are also widespread and these probes are also available. Some probes can be switched from one to the other or used on both. The probe is connected to the power supply of the logic being probed. A light in the probe indicates the state of the voltage attached to the probe tip. A high voltage (+5 volts for TTL, +15 for CMOS) lights the light (usually an LED). A low voltage puts the light out. A blinking light indicates pulses on the point.

Questions 6 & 7 concern themselves with testing a logic circuit and a logic probe. If AND gate 1 was operating properly the probe should blink at F with the conditions stated. But, if the input to OR gate 3 is shorted to ground the probe at F would not blink. Hence, even though answer a is correct, answer b is better because it is more inclusive.

Suppose that the probe blinked on point A and blinked on point B. Would a

"0" indication at F mean the same thing, that 1 or 2 is bad? Not necessarily. Remember that to get a "1" out of an AND gate all inputs must be "1" simultaneously. The probe might blink at A and at B and still not have a "1" on each at the same time. On the other hand, if C and D inputs to the OR gate blink, point G should also cause the light to blink. But, if the logic level at C and the logic level at D never go to the zero logic level at the same time, then G could indicate a "1" at all times. Therefore, answer b is the correct response in Q8.

Question 9 points out another area where care must be exercised not to jump to false conclusions. If the input from line F to the gate, inside the IC chip, is open, a logic probe could easily blink at all three points: F, G, and H; and the gate would be bad. Answer d is the correct response.

A logic pulser is also a probe available for troubleshooting logic. This device sends pulses into the point attached to the probe. This is useful in toggling gates on or off. It is usually used in conjunction with a logic probe, the probe detecting the output of the gate being switched or toggled.

In Q10 the transistor should stay off as long as B is grounded, answer a. However, if B is lifted from ground a pulse should be indicated by the logic probe at the collector. Most logic probes will blink whether the input to the probe is low most of the time, or high most of the time.

Logic analyzers have 8, 16, or 32 input lines. The CRT or other display shows the activity on each of these input lines. Some displays can be selected for binary, hex, or octal. If a CRT is used, timing and memory map modes are also available. The answer to Q10 is d. In any of these modes of operation, it is necessary to know the proper sequence, or pattern of data on the many lines, to interpret the results.

A good deal of computer troubleshooting is accomplished without the need for electronics or programming knowledge. This is made possible by relying on diagnostic routines available on most manufacturer's computers, printers, etc. Most machines have a self-test incorporated in the "turn on" procedure. When a trouble spot is isolated to a subassembly or PC board, that board or subassembly is exchanged and hopefully the machine operates. Most manufacturers have an exchange program for these boards or subassemblies.

Unfortunately, this kind of procedure ties technicians closely to the manufacturer. And what happens when the manufacturer no longer supports the exchange program on a particular unit? Then the technician needs to be able to isolate problems down to the component level to be able to repair that particular unit.

Common computer equipment problems

By far the largest group of problems with computer equipment is mechanical, or related to mechanical movement. For example, floppy-disk drives cause a good deal of trouble by becoming misaligned; that is, the record/play-back head becomes misadjusted from wear, vibration, or impact. Printers cause a lot of problems with platen drive, carriage drive, and hammer drive problems. The keyboard might also have keys that stick, or fail to make contact. Keyboards are also susceptible to foreign material problems such as coffee and cola spills.

Another common mechanical problem involves connectors. Many computers are repaired by removing and reinserting IC chips (socket mounted) and connectors. Special cleaners can be purchased for connectors.

ESD and safety

Electrostatic Discharge is an important subject. As the density of components on IC chips increases and with the increase in usage of MOSFET chips, static damage is an ever present danger. Computers and other digital equipment should be worked on at a static-free workstation.

Many technicians say they never had problems with static and hence didn't need wrist straps, static pads, nickel-plated bags, or antistatic packing material. The trouble is that it is hard to point to a problem caused by ESD damage. No one knows for sure. But companies that take steps to protect components from ESD see call-backs reduced, and manufacturers see reliability increases that are sometimes dramatic.

Additional computer terminology

Below are some acronyms that technicians should become familiar with:

BIOS Basic input-output system.
CGA Color graphics array.
DMA Controller Direct memory access controller.
EGA Enhanced graphics adapter.
EMS Expanded memory specifications.
MDA Monochrome display adapter.
SIMM Single in-line memory module.
VGA Video graphics array.

Additional reading

Coffron, James W.: *Practical Troubleshooting Technique for Microprocessor Systems*, Prentice-Hall, 1981.

Inbody, Don: *Principles and Practice of Digital ICs and LEDs*, TAB Books.

Lenk, John D.: *Handbook of Practical Microprocessor Troubleshooting*, Reston Publishing Co., 1979.

Namgostar, M.: *Digital Equipment Troubleshooting*, Reston Publishing Co., 1977.

Osborne, Adam: *An Introduction to Microcomputers, Volume 0 the Beginner's Book*, 2nd Ed., Osborne & Associates, 1979.

Paynter, Robert T.: *Microcomputer Operation, Troubleshooting and Repair*, Prentice-Hall, 1986.

Short, Kenneth L.: *Microprocessor and Programmed Logic*, Prentice-Hall, 1981.

Slater, Michael and Barry Bronsen: *Practical Microprocessors*, Hewlett-Packard, 1979.

15
The Industrial option

Quiz

Q1. In Fig. 15-1 the bulb will light as long as SW1 is making contact and will not light when SW1 is not making contact.

a. True
b. False

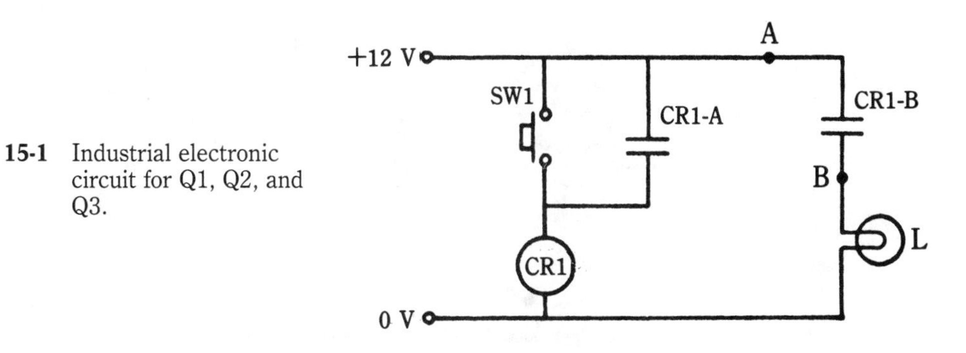

15-1 Industrial electronic circuit for Q1, Q2, and Q3.

Q2. When the voltage at A is equal to the voltage at B the light is out.

a. True
b. False

Q3. The purpose of the CR1-A relay contacts is to keep the relay ON after the SW1 switch is released.

a. True
b. False

Q4. Match the following industrial component symbols:

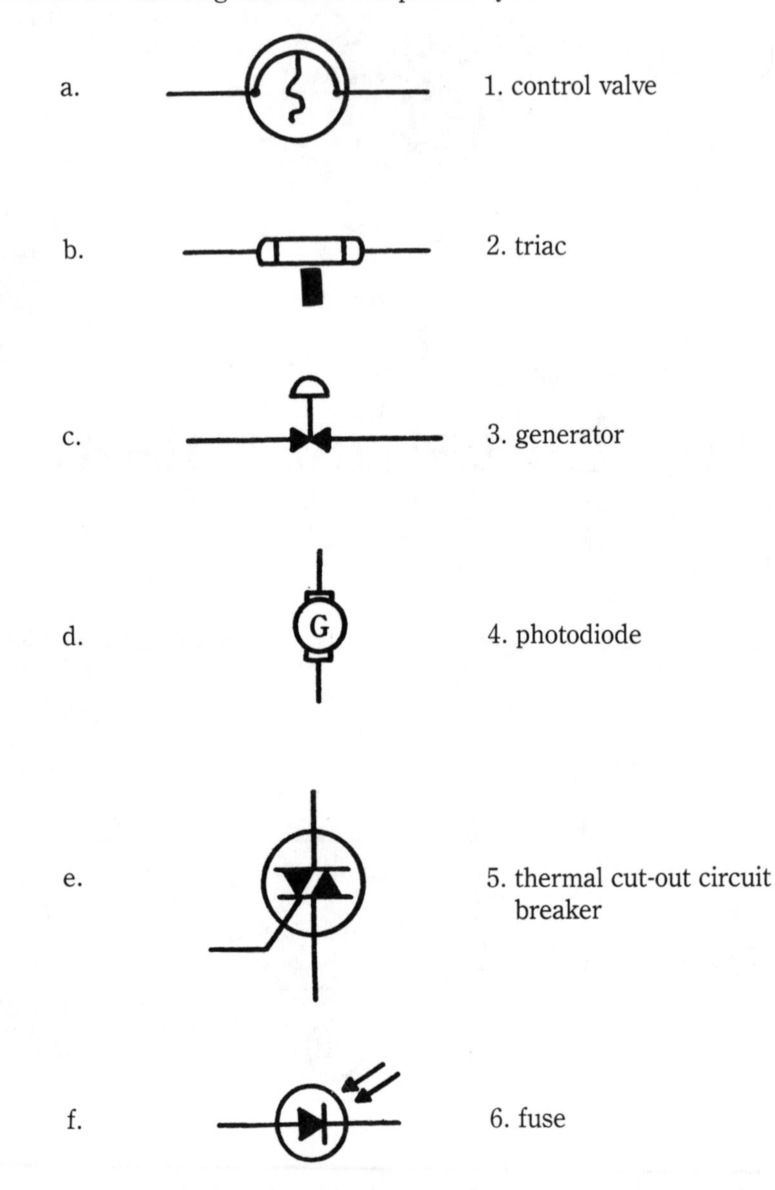

Industrial electronics

The SW1 switch in Q1 makes contact, which causes the relay contacts to close—both CR1-A and CR1-B. As long as CR1-A is closed, it will keep current flowing through CR1 and thus keep the contacts closed. Once the circuit is opened by removing the 12 V (perhaps by turning off an on/off switch), the relay de-energizes, the contacts open up, and the light goes off. Part of the difficulty with industrial electronics is that unless you work with it regularly you might not recognize some of the symbology and notations. Relay contacts look like capacitors in other symbology and the relay does not show the "coil of wire" symbol familiar to other areas of electronics.

In Q2 (Fig. 15-1), when A and B are equal voltages the CR1-B contacts must be closed and the lamp is lit unless the 12-V supply is missing. Assuming the 12 V is as shown, with the contacts open, A will be + 12 V with respect to 0-V reference. B will be at the potential of the minus side of the 12-V supply or 0 V.

Question 3 explains the purpose of the CR1-A contacts. This allows SW1 to be a momentary, low-power contact switch, rather than a toggle or rotary mechanical switch. In high-power circuits, SW1 could be a smaller control-panel button switch while the relay could be located elsewhere closer to the main power and the activated device, perhaps a light bulb or motor.

The symbol in Q4 a is that of a thermal cutout type of circuit breaker. This is one of several fuse/breaker types found in industrial electric/electronic drawings. The ordinary current overload circuit breaker is often presented as in Fig. 15-2.

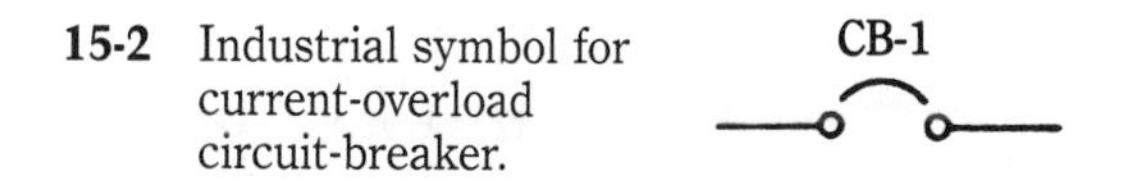

15-2 Industrial symbol for current-overload circuit-breaker.

The thermal cutout depicted in a of Q4 shows the added heating element symbol, which confirms that it is heat that causes the circuit breaker to open. The symbol in Q4 b is a fuse. In industrial work this style of fuse is a carry-over from the physical appearance of the cartridge fuses so common in electrical work. Two other symbols are also commonly used today for fuses (Fig. 15-3). F1 is normally a glass-enclosed wire while F2 is a fusible resistor, usually wire-wound.

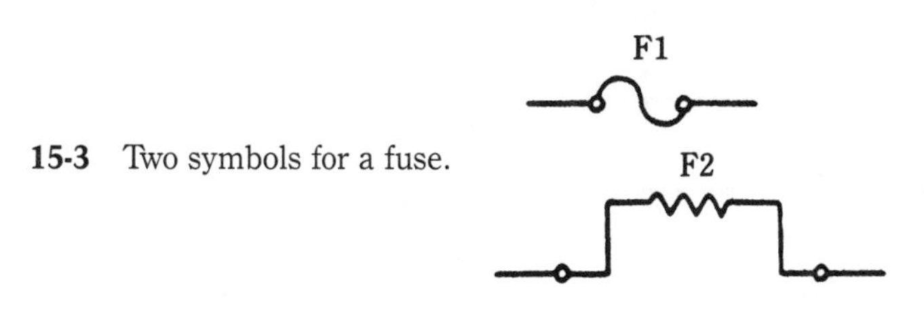

15-3 Two symbols for a fuse.

The symbol in Q4 c is a control valve such as that used in the process-control industry to change the flow of steam, air, or fluids. Today such devices are directly controlled by computers that sample many important locations in manufacturing processes. Petroleum refining, insulation manufacturing, paper making, and many others use control devices. Pressures, levels, and temperatures are monitored and compared in these systems. Transducers like c are then caused to open or close in exact amounts necessary to optimize the operation for maximum efficiency and safety.

The symbol for a generator is shown in Q4 d. Because alternators, starters and motors are similar in appearance, the G is used to positively identify that the device is a generator.

e in Q4 is a triac. A single SCR can be used in a dc circuit for many purposes.

An SCR used in an ac circuit cuts off 1/2 of the waveform, as in Fig. 15-4b. A triac behaves like back-to-back SCRs connected as in Fig. 15-5. Triacs and SCRs can be used in heavy current circuits while being controlled by a gate signal much smaller than the controlled current.

Item f is the symbol for a photodiode in Q4. The photodiode must have a trigger. This comes as an outside light source. Arrows pointing *in* indicate a semiconductor device that is activated by light. Arrows pointing *out* indicate the device produces light.

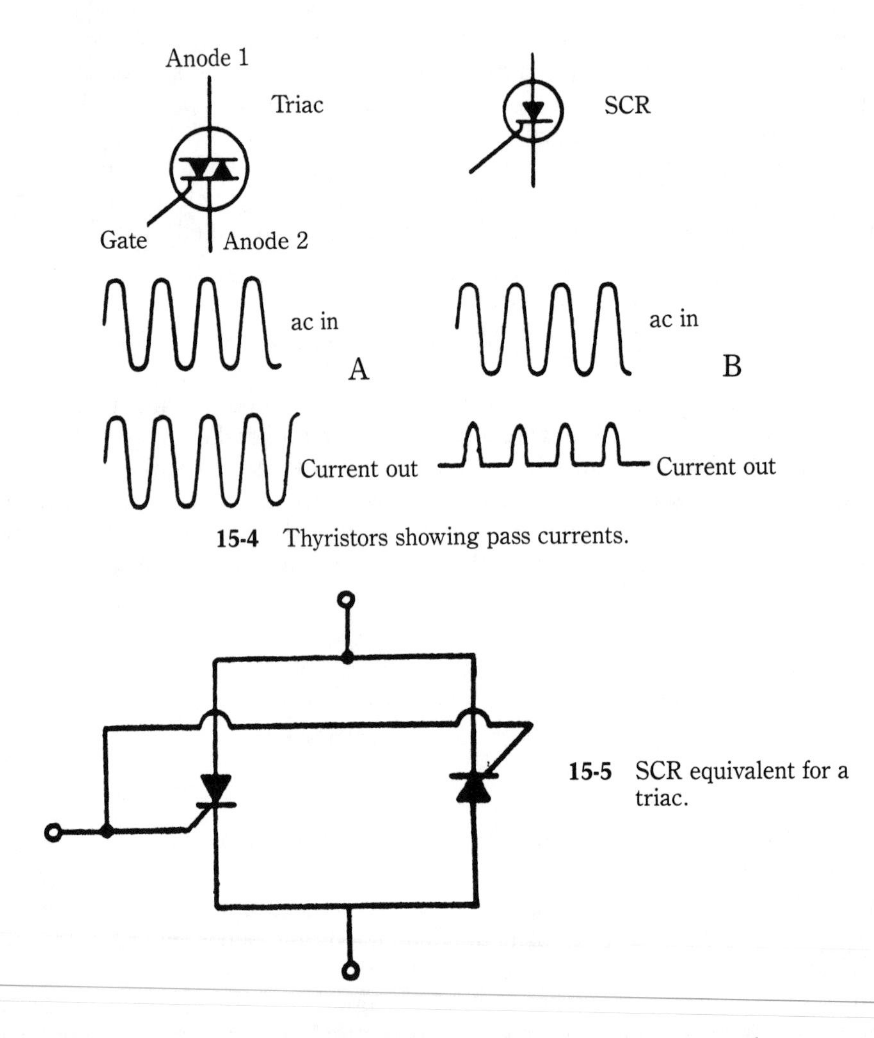

15-4 Thyristors showing pass currents.

15-5 SCR equivalent for a triac.

Additional questions on the Industrial option

Q5. The function of a tachometer is:

a. to detect errors in a speedometer reading.
b. to transmit velocity information.

c. to report increases or decreases in speed.
d. to increase a vehicle's acceleration rate.

Q6. The key signal in a closed-loop process controller is generally considered to be the differential signal.

a. True
b. False

Q7. Frequency counters, given the same level of signal, are most accurate at low frequencies.

a. True
b. False

Q8. Current ratings on SCRs are usually listed as:

a. average.
b. rms.
c. total.
d. instantaneous.

Q9. Base speed of a dc motor occurs at:

a. full armature voltage and low field current.
b. low armature voltage and full field current.
c. low armature voltage and low field current.
d. full armature voltage and full field voltage.

Q10. Viewing a 14-pin DIP IC from the top, the pin identification numbers begin at the notch on one end and are counted clockwise from there.

a. True
b. False

Q11. In Fig. 15-6 the controller marking SP stands for:

a. system process.
b. set point.
c. steam power.
d. serial pulses.

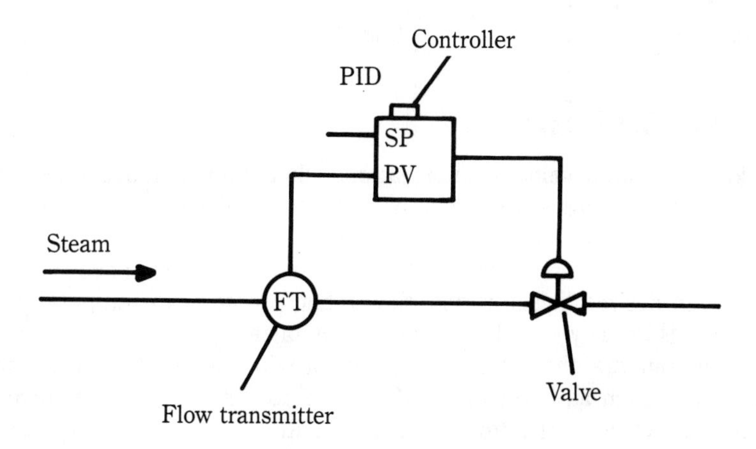

15-6 Process Control system.

Q12. A magnitude tachometer is really a:

a. dc generator.
b. squirrel.
c. rotating transformer.
d. None of the above.

Q13. As temperature changes, an RTD will change its:

a. voltage output.
b. resistance.
c. color.

Q14. An LVDT is:

a. low-voltage dielectric transducer.
b. linear-voltage display television.
c. linear-variable differential transformer.

Q15. Do strain gauges measure resistance changes?

a. No; they measure weight changes.
b. Yes; they measure torque, force, and weight changes.

Q16. In Fig. 15-7, when VOM resistance measurements are taken on the transistor, they measure as shown. We can conclude:

a. it is an NPN.
b. it is a PNP.
c. the transistor is bad.
d. lead B is the collector.

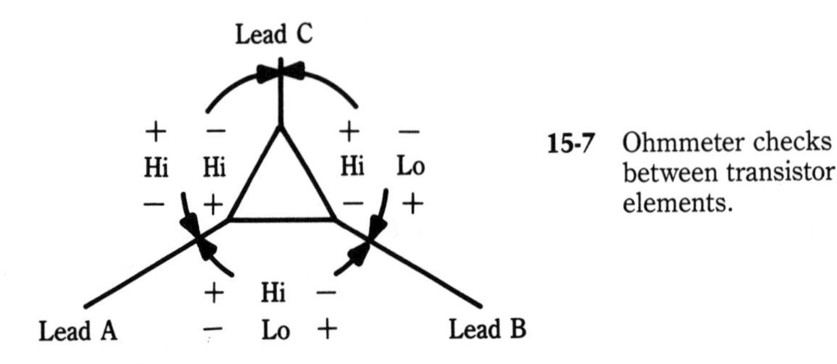

15-7 Ohmmeter checks between transistor elements.

Quiz explanation

A tachometer transmits velocity information. That information can be used to detect errors, speed increases or decreases, or to indicate if a vehicle's rate of acceleration has increased. The correct answer is b.

The closed-loop process control system has several signals. These signals are used to maintain desired levels, ratios, flow rates, pressure, or temperature. The feedback signal is the key signal, therefore Q6 is false.

Frequency counters characteristically have a wide range of input sensitivity. Low frequencies, such as 5 Hz to 80 MHz, typically require a 25-millivolt rms input. Signals above 600 MHz (to 1300 MHz) require six times as large an input signal (150 mV rms) to provide accurate readouts. The best answer to Q7 then is true. Another interesting characteristic is the wide range of maximum input voltages allowed. Some allow 150 volts rms at frequencies up to 10 kHz. This maximum allowable voltage input drops to 50 V rms at 10 kHz and only 1 V rms at 80 MHz. The best answer then is a.

Question 8 notes that SCRs are listed for their average current rating.

Question 9 wants you to know the base speed of a dc motor that occurs when full armature and full field voltage is attained.

By now you should know the pin numbering system universally used with integrated circuits (ICs). From the notch on one end of the DIP (Dual In-line Package), looking at the top of the chip, the number 1 pin is to the left. Numbering increases to the bottom, then back up the right side (Fig. 15-8).

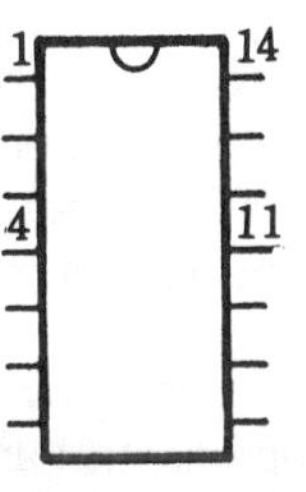

15-8 Top view of Integrated Circuit chip.

Question 11 asks about the term SP as used in process control work. That SP stands for Set Point; thus, b is correct.

Of the two types of tachometers (frequency and magnitude), the magnitude tach is basically a dc generator, answer a.

An RTD is a resistance temperature device that has a positive temperature coefficient, increasing its internal resistance as its temperature increases.

An LVDT is a linear variable differential transformer, used to provide a voltage proportional to a linear movement of a shaft. The answer to Q14 is thus c.

Strain gauges in Q15 measure resistance changes. They are ordinarily connected in adjacent legs of a bridge circuit. They can measure torque, force, and weight changes.

The last question requires you to determine the type of a transistor, as well as its lead connections. If the transistor is not defective, by placing the positive lead of your ohmmeter on the base, and the black or negative lead on an NPN transistor's emitter, you should read a low resistance. You have forward biased that diode-like junction and turned it on.

If you forward bias the collector-base junction, you get the same result; a low reading when the collector has the negative lead connected—high when reversed. Because connecting the positive lead to the collector and the negative to the base won't turn on any junction, that reading will be high and so will the reverse lead position (Fig. 15-9).

By understanding this concept the type of transistor can be recognized and the collector, emitter, and base can be identified. In Fig. 15-7 notice that either position between leads A and C gives a high reading. Therefore one must be the collector and the other the emitter. But which is which? To find out, see which way the

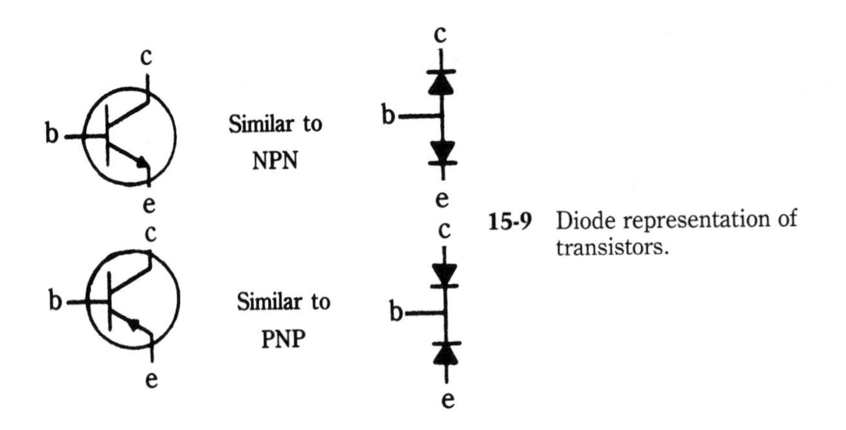

15-9 Diode representation of transistors.

other connections read high or low. In this case + on B and – on C gives a low reading. The reverse lead position gives a high reading. Because C and A read high both ways, this transistor is probably an NPN and C is the emitter. But let's make sure.

Between A and B we can turn on the junction by placing the red lead on B. Looking at the diode representation in Fig. 15-9, the collector/base junction should check low with + on the base. Therefore B is not the collector, but the base. The transistor checks good and it is really an NPN (Figs. 15-10 and 15-11).

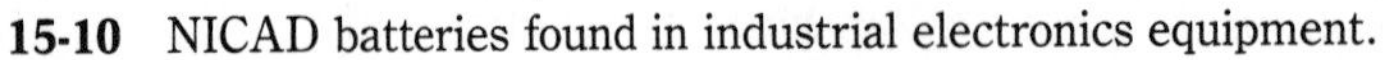

15-10 NICAD batteries found in industrial electronics equipment.

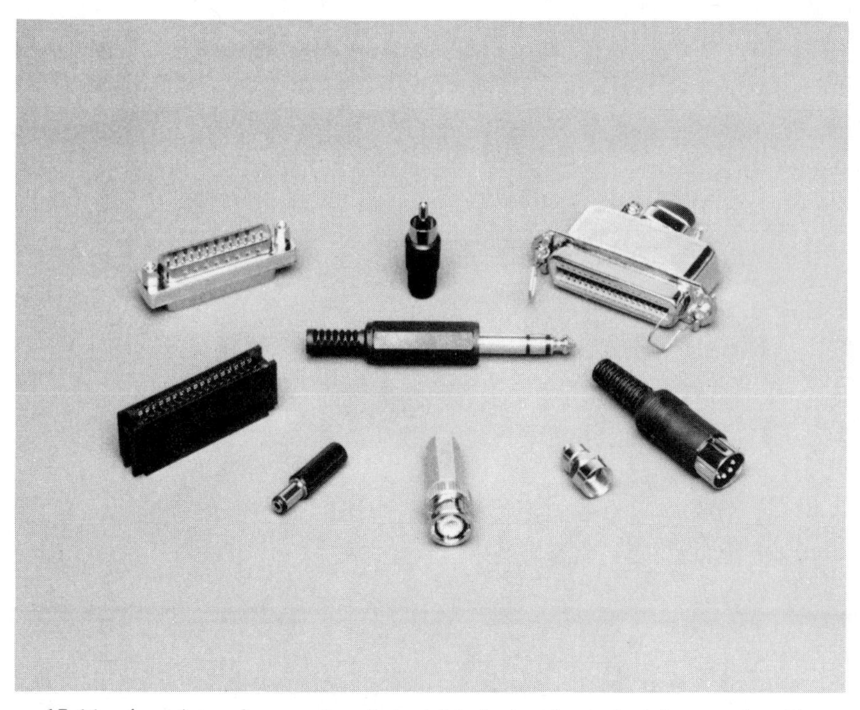

15-11 A variety of connectors industrial electronics technicians work with.

16
The Biomedical option

Quiz

Q1. The calibration button on an ECG (or EKG) machine does what?

a. produces a 1 microvolt pulse.
b. produces a 1 millivolt pulse.
c. produces a 1 microvolt dc level.
d. produces a 1 millivolt dc level.

Q2. The heart produces a small bioelectric potential due to:

a. friction of the blood through the heart arteries.
b. sodium-potassium imbalance in heart cells.
c. neutralized cardiac arhythmias.
d. the Doppler shift principle.

Q3. More than one lead is used in connecting a patient to the ECG machine. Why?

a. Several leads assure at least one positive surface connection.
b. It is easy to recognize a heart disease problem if all leads do not produce identical waveforms.
c. Half of the leads respond to P waves, the other half to augmented waves.
d. Each lead gives information from a particular portion of the heart muscle.

Q4. In ECG recordings, artifacts are:

a. extraneous signals appearing on the ECG tracing.
b. the "optimal" sample of a perfect reading against which the actual ECG is compared.
c. the previous ECG reading from the same patient.

Q5. While dangerous current levels differ slightly from patient to patient, the amount that can cause painful electrical shock is:

a. less than 900 microamps.
b. less than 5 milliamps.
c. less than 20 milliamps.
d. less than 300 milliamps.

Q6. The difference in contact resistance between dry and damp skin can be a ratio of:

a. 2 to 1
b. 10 to 1
c. 100 to 1
d. 1000 to 1

Q7. While the skin contact differences in Q6 are extremely important, medical patients frequently have electrical apparatus connected through pierced skin, catheterization, etc. What level of leakage current through the body is considered unsafe?

a. 2 microamps
b. 20 microamps
c. 100 microamps
d. 1 milliamp

Q8. The third wire in a power cord has the purpose of:

a. carrying leakage currents to ground.
b. providing an alternate neutral path should there be a defect in the neutral wire.
c. shorting the HOT side of the line to ground in case of overload.
d. shorting the HOT side of the line to neutral in case of overload.

Q9. Brain signals are recorded with a/an:

a. electro myogram.
b. electro encephalogram.
c. plethymographic transducer.
d. systole lamp or pulsometer.

Q10. Differential amplifiers are used in input circuits of medical electronics equipment because they:

a. have better high-frequency characteristics than most other amplifiers.
b. amplify common mode signals.
c. have higher voltage gain than other amplifiers.
d. reject common mode signals.

Q11. X-ray tubes use anode potentials of:

a. 400 to 550 volts.
b. 4000 to 5500 volts.
c. 40,000 to 100,000 volts.
d. over 150,000 volts.

Q12. In Fig. 16-1 the diodes will, in the charge position:

a. charge C1 to 110 volts.
b. charge C1 to over 5000 volts.
c. charge C1 to 200 volts.
d. charge C1 to 24 volts.

Q13. In Fig. 16-1 the meter is calibrated to read:

a. discharge current
b. volts
c. resistance
d. energy in watt-seconds

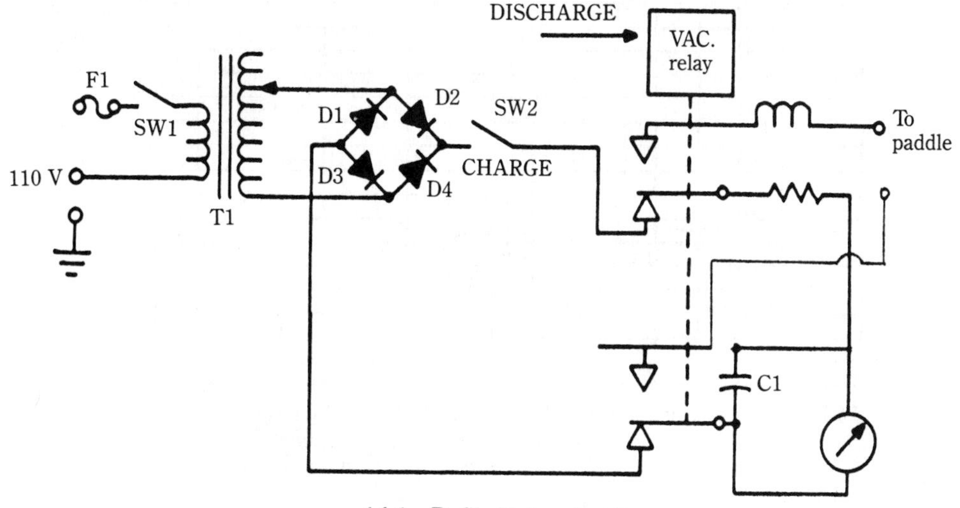

16-1 Defibrillator circuitry.

Q14. Medical oscilloscopes normally are able to reproduce higher frequencies than ordinary service oscilloscopes.

a. True
b. False

Q15. Electrosurgery machines usually use which of the following?

a. foot switch
b. stylus damper
c. pressure transducer
d. watt-second meter

Quiz explanation

Standard practice is for the calibration button on an ECG machine to produce a 1 millivolt pulse; thus the answer to Q1 is a.

A section of a heart muscle will continue to pulse when cut away from the main portion. The different mix of sodium and potassium on one side of a heart cell versus the other causes a battery-like buildup of electrical charge near 100 millivolts. This difference in potential can be measured at different points on the body.

More than one lead is connected to the patient. The different ECG readings obtained are then used to diagnose heart conditions. Figure 16-2 shows a typical ECG response waveform. The correct answer to Q3 is d.

Artifacts are unwanted signals riding on the ECG recording. These can be "hum" interference or pickup; they can be noise pickup or other erroneous information. The answer to Q4 is thus a, extraneous signals appearing on the ECG tracing.

The amount of current expected to cause a painful shock is considered to be about 15 milliamps. The correct answer to Q5 is c; however, less than 300 mA is also a pretty good answer. At about 15 milliamps, a person still can pull away from

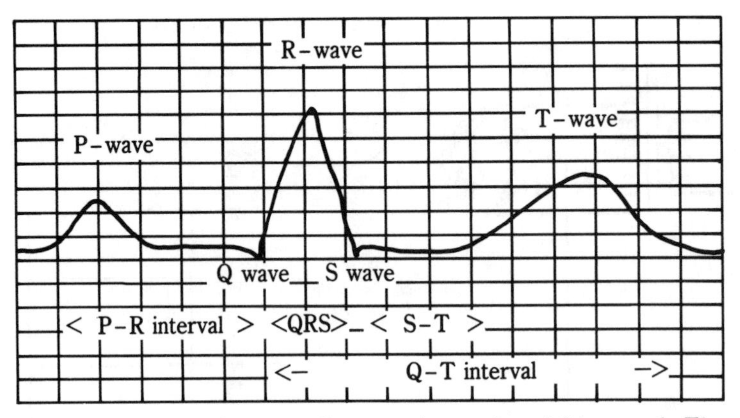

16-2 The width of one small square is equal to 0.04 second. The height of one small square is equal to 1 millimeter. The above complex shows how a normal cardiac cycle would look when graphed.

the contact. Above that, the shock is painful and the person might not be able to pull away or might do physical harm to himself due to mechanical reaction to the shock.

In Q6, the subject of contact resistance differences is addressed. Skin resistance is much higher when dry, about 1000 times as high. The correct answer to Q6 is d.

Much greater danger from defective electrical equipment exists for patients who have equipment connected to them. Extreme care is needed here as the resistance ordinarily provided by the skin is no longer there. Where an electrical burn on the skin might build up an even higher resistance, lowering the shock potential, a medical patient connected to wires, probes, and sensors has no such safety factor. It takes only 20 microamps to be unsafe. Answer b is correct for Q7.

The third wire in a power cord (usually green) carries leakage currents to ground. That is all; thus, answer a is correct for Q9.

EEG, or electro encephalogram, is the name for the device used to record brain wave information.

Differential amplifiers have the ability to reject common-mode signals that aren't "common." Common-mode signals such as you might expect from two different grounded chassis might differ slightly, or if there is a defect, they might differ greatly. Therefore common-mode problems are potential dangers in all electrical equipment. The correct answer for Q10 is d.

X-ray tubes operate by using high accelerating voltages to drive a photon into the nucleus of an atom where it collides with an electron, which gives the electron energy to travel away in a beam along with other electrons. This beam passes through the patient and then onto a fluoroscope or onto film. The correct answer to Q11 is c.

The diodes in Fig. 16-1 will charge the capacitor C1 to the peak value of the transformer secondary tap selected. This can be near 5000 volts. The actual charge from the paddles to the patiènt is less than a tenth of that amount. The correct answer to Q12 is b.

The meter in Fig. 16-1 is calibrated to read discharge current. The best answer for Q13 is a.

Medical oscilloscopes rarely need to reproduce high frequencies. More often, very low frequencies, such as heart rates at less than 1 per second and other body rates, are viewed. The correct response to Q14 is b.

The last question asks if you know that an extra hand is helpful in surgery. The foot switch gives the surgeon that help by turning the electrosurgical stylus on or off as needed while both hands are busy with other requirements.

17 Video distribution option

Quiz

Q1. The polar pattern in Fig. 17-1 is most likely for:

a. a vertical dipole.
b. a UHF corner reflector type TV antenna.
c. a horizontal dipole.
d. a conical.

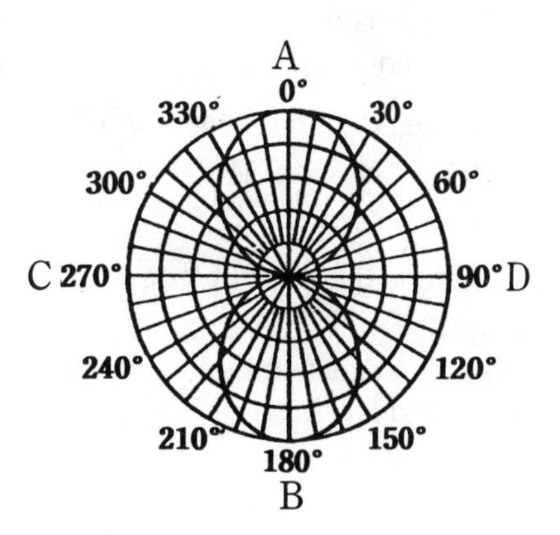

17-1 Polar pattern for Q1. The figure "8" between A and B is the antenna's response graph.

Q2. In Fig. 17-2 the name for the element marked C is:

a. director
b. reflector
c. active element
d. boom

Q3. The antenna in Fig. 17-2 would be called a log-periodic.

a. True
b. False

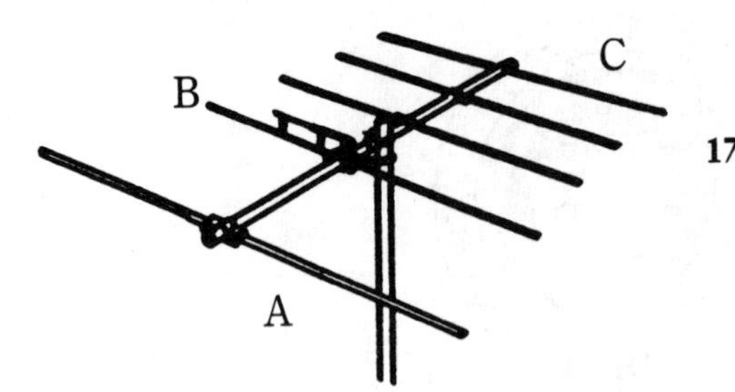

17-2 Antenna with the three basic elements identified A, B, or C.

Q4. The strength of an RF signal that was previously measured as 1000 microvolts has now been reduced to only 500 microvolts. This represents what amount of loss in dBs?

a. 500 c. 6
b. 10 d. 3

Q5. A 2-way signal splitter output port ordinarily loses 50 percent of the input signal.

a. True
b. False

Q6. Match the following frequency bands:

a. AM Broadcast Radio	1. 88 – 108 MHz
b. FM Broadcast Radio	2. 3.7 – 4.2 GHz
c. CB Radio	3. 54 to 88 MHz
d. UHF TV	4. 27 MHz
e. TV superband	5. 535 – 1605 kHz
f. audio frequencies	6. 20 Hz – 20 kHz

Q7. Frequencies ranging from 300 MHz to 3000 MHz are identified as:

a. UHF c. SHF
b. VHF d. EHF

Q8. How wide is the band of frequencies used by the following?

a. audio	1. 10 kHz
b. AM Broadcast station	2. 200 kHz
c. FM Broadcast station	3. 20,000 Hz
d. TV Channel	4. 6 MHz

Q9. If a single crystal generates a frequency of 1 MHz, that signal is not modulated (or impressed with any audio or video information) and that signal is broadcast, what bandwidth will it be using?

a. 1 MHz c. 2 MHz
b. 1 Hz d. none

Q10. In Fig. 17-3 what is the total wire loss?

a. 4 dB

b. 24 dB
c. 32 dB

Q11. In Fig. 17-3 is the signal level out of the tap at X large enough?
a. No, there is no signal out.
b. No, tapoffs should have at least 10 dB output.
c. Yes, the signal is perfect for a TV set.
d. Yes, the signal is sufficient to supply several TVs.

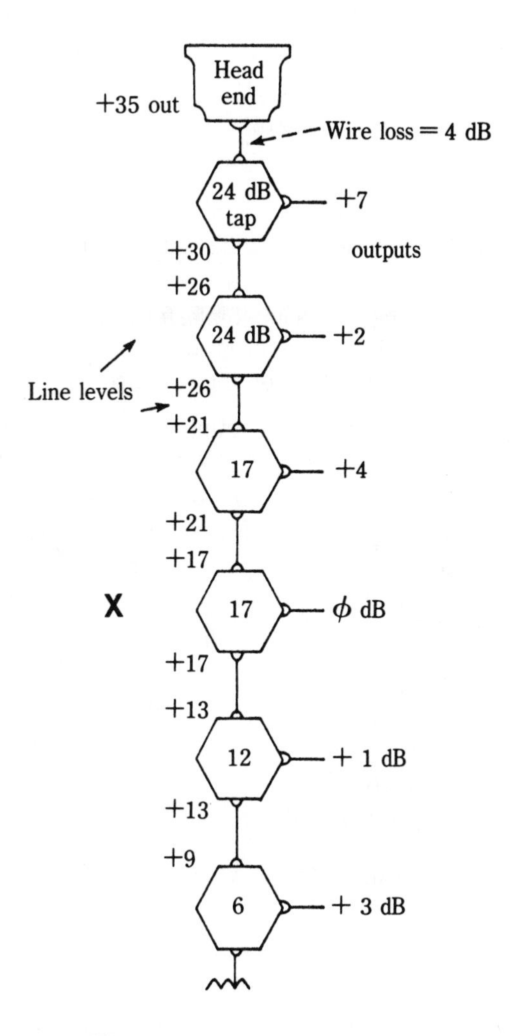

17-3 Wire loss calculations in an MATV system.

Q12. The tap losses shown (24 dB, 17, 12, and 6) are called

a. insertion losses.
b. isolation losses.
c. line losses.

Q13. Match the following head-end equipment to its use:

a. modulator	1. boost signal down line
b. strip amplifier	2. ch 30 to ch 12
c. pre-amplifier	3. off-air channel
d. line amplifier	4. VCR channel
e. U/V converter	5. antenna signal booster

Q14. Cable TV systems use dc power on the trunk lines to power trunk amplifiers and line extenders.

a. True
b. False

Q15. Interference from a distant TV channel of the same frequency as the desired channel is called:

a. adjacent channel interference
b. co-channel interference
c. intermodulation interference

Q16. Most antenna pre-amplifiers use the same coaxial or twin-lead transmission line to supply the TV-radio frequencies as well as to supply the mast-mounted amplifier with ac power.

a. The same wire is used but the power is dc, to eliminate hum in the signal.
b. The same wire is used for transferring the channels down to the TV, but signal rectification is used to power the pre-amplifier.
c. True.

Q17. Match the following:

a. 17 dB loss	1. attenuator
b. 7 dB loss	2. 4-way splitter
c. 24 dB gain	3. amplifier
d. 8 dB loss on low VHF/ 0 dB loss on high VHF	4. tap off
e. separate U and V signals	5. line compensator
f. reduce all signals 10 dB	6. bandsplitter

Q18. Satellite receivers, VCRs, and CD players supply approximately what signal level out?

a. 100 millivolts	c. 2000 microvolts
b. 1000 millivolts	d. 10,000 microvolts

Q19. At 400 MHz, how much more wire loss is there in RG-59 copper-braid coax than in RG-6 foam coax?

a. 5 percent c. 25 percent
b. 15 percent d. 50 percent

Q20. A signal level meter or Field Strength Meter will read what range of TV signals without external attenuators?

a. −40 to +40 dB c. 0 – 100 microvolts
b. 0 – 100 dB d. 0 – 1000 microvolts

Quiz explanation

This polar pattern in Fig. 17-1 shows that the antenna receives signals just as well from the forward direction as from the rear. This means that the antenna has no directors or reflectors. Therefore it must be a single dipole horizontally mounted antenna. A UHF corner reflector has good forward gain, but the reflector prevents rearward signals from being received. A vertical dipole, such as an auto's CB antenna, has a 360-degree horizontal pattern or a circular pattern. A conical antenna has a much wider forward lobe than that in the figure, as well as a much smaller rearward lobe.

The element marked C is a director in Fig. 17-2. The directors are at the front end of the antenna. The reflector on this YAGI antenna is at A, the rear. The active element that is connected to the transmission line is B. These YAGI cut-channel antennas are recognizable due to each element being of similar length. They have high gain, those with the most elements approaching 18 dB, as compared with a single half-wave dipole. The polar pattern of a YAGI is nearly all forward, with very little reception from the 45-degree angles, and practically none from the rear.

Question 3 asks if the antenna could be called a log-periodic. The answer is no. In Fig. 17-4 you see a log-periodic style. Notice it is an all-channel antenna, because it has different lengths on the various elements. Each set of elements is cross-linked to the next set behind or in front of it. Such antennas use the small front elements to receive the high channels, but to also act as directors for the channels resonant to the larger elements behind.

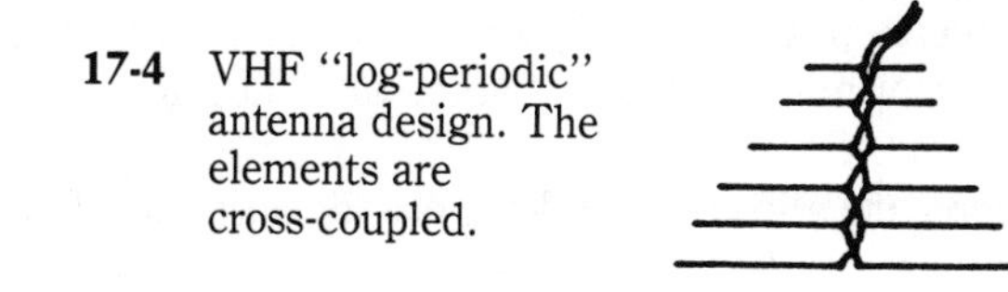

17-4 VHF "log-periodic" antenna design. The elements are cross-coupled.

Question 4 asks if you know what the loss is to a signal voltage that is reduced by 50 percent. This is something you must know. When a signal VOLTAGE increases by 100 percent, it is increased by 6 dB. When it is decreased by 50 percent, it is a decrease of 6 dB. When dealing with power or watts, the dBs are only 3 in each case, so don't confuse voltage dBs with power dBs.

Although you might think a splitter that divides the signal in two reduces the signal by 50 percent (or 6 dB), it doesn't. Instead of reducing the signal by 50 percent, it is only reduced by about 37 percent. This is due to the selection of the transformer ratio used in coupling the input to the two outputs. The loss in a 2-way splitter is only 3.5 dB, rather than 6. You can figure out the loss in other splitters by understanding that internally most splitters are made up of 2-way splitters. A 4-way splitter is really a 2-way inside, with both ports of the two way connected to two more 2-ways; thus, four outputs. The loss is 3.5 dB through the first 2-way splitter, plus the 3.5-dB loss through the second, for a total of 7 dB loss on a 4-way splitter.

Matching the frequency bands in Q6:

a = 5
b = 1
c = 4
d = no match
e = no match
f = 6

The CET exam doesn't have a set of matches that have no correct answer for some of the choices. In Q6, the right choice for UHF would be 470 to 806 MHz. The right choice for TV superband would be 216 to 276 MHz (channels J through W). The choice of 3.7 to 4.2 GHz is for the 4 GHz C-Band Satellite channels. The VHF-low band is 54 to 88 MHz.

In Q7, the UHF band is 300 to 3000 MHz. In communications radio, techs usually identify the UHF business radio band as that narrow band around 800/900 MHz. This is UHF, but it isn't the total spectrum identified by the FCC and the communications world as UHF. The VHF band is 30 to 300 MHz and the Super High Frequency band is 3000 MHz to 30,000 MHz, with EHF, or Extremely High Frequencies above that, to 300 GHz.

To help understand the relative size of the bands of frequencies used by each service, Q8 asks if you know how much frequency band or spectrum is used by four common segments: Audio frequencies (if your hearing is good) might be from 20 Hz to 20 kHz. Some books show it as 16 Hz to 16,000 Hz. Some audio equipment boasts its frequency response is from 5 Hz to 25,000 Hz. Even though most people can't hear frequencies much above 15 kHz, it is claimed that those sounds above the ordinary range can be discerned by some and therefore they add to the "presence" sensation of high-fidelity music.

AM broadcast stations are allowed a 10 kHz broadcast band. This means the signal can only be modulated with as much as a 5 kHz audio signal. This might change soon as AM radio is being allowed to enhance its signals in a manner that will improve the frequency spectrum and to also include stereo AM radio.

FM broadcast stations use up to 200 kHz of spectrum to provide frequency-modulated audio. TV channels gobble up 6 MHz of signal. Channel 2 runs from 54 MHz to 60 MHz. A single TV channel uses about six times the frequency spectrum as is used by the entire AM radio band, which can include dozens of local stations.

In Q9 the concept of bandwidth is continued, because it seems to confuse many technicians. Broadcasting technology and modulation of a signal enter into the matter. This question asks what happens if there is no modulation? If a single frequency is being transmitted, what happens? That single frequency is transmitted with no sidebands, all by itself. With no modulation of the 1 MHz signal, that is all that is being transmitted, that single frequency of 1 MHz. Therefore the bandwidth of that station or channel is only one cycle, or 1 Hz wide. You would find nothing on either side of it, either higher or lower. If you talk into a microphone that is connected to a modulation circuit in the transmitter broadcasting this 1 MHz frequency, the various frequencies in your voice cause the original 1 MHz frequency to produce sideband frequencies. This broadens the spectrum used by the station, perhaps to + or − 20 kHz from that 1 MHz frequency, if AM, and much wider than that if FM.

In Q10 you are asked to look at the MATV circuit and decide how much wire loss there might be between the head end and the terminated end. The drawing indicated between the head end and the first tapoff there is a 4 dB loss because of wire loss. From there you can see from the signal levels shown that there is a 4 dB loss between each tap; therefore, the total wire loss is 24 dB. There might be some loss between the last tap and the terminator, but that terminator should be screwed onto the tap itself; thus, there would be no additional wire loss.

In Q11, the signal out of the tap at X is 0 dB, which is 1000 microvolts, perfect for a TV input. While cable companies might plan to serve any home with 6 – 10 dB of signal, not knowing that more than one TV is to be hooked to X leaves it as sufficient. Because most homes have more than one TV, a 6 – 10 dB signal is becoming required. But the answer to Q11 is c.

In Q12, tap losses as shown, the loss of signal from the feed-through input and output of the tap to the tap-off port, are indicated in drawings and on the tap itself. This is called isolation loss, although another name would be easier to remember, like tap-off loss. The loss in a tap between the input and output to the next tap, or through-loss, is called the insertion loss. Just about any piece of hardware you put a signal into has some type of insertion loss, even F connectors and F-71 and F-81 splices. Line losses are considered to be the wire loss.

Matching the Q13 equipment:

a = 4
b = 3
c = 5
d = 1
e = 2

Two types of amplifiers are used in SMATV, MATV, and cable head-ends. One is a modulator. This must have a video and an audio input signal. A VCR's audio and video outputs can be input into a modulator. That modulator takes the input video and audio and impresses these two signals onto a carrier signal at the channel frequency you choose.

A strip amplifier takes an off-air signal and merely builds it up to sufficient size to be used by the system it is to be included in. 30 dB to 50 dB are common strip amp and modulator output levels.

The pre-amplifier is used on the antenna mast itself near the antenna arrays to amplify the signals prior to any line losses or noise pickup. These can be 17 to 34 dB in size.

Line amps are usually used after the signals have been processed and placed out on the trunk line(s). They usually require a larger input than pre-amplifiers, on the order of 10 dB. They build up the signals that have been attenuated through taps, splitters, and coax lines and are placed in the line at intervals.

U/V converters (as well as V/V and V/U converters) are used to change the channel frequency. You might want to make a UHF channel into a VHF channel, because UHF loses much more signal in long wire runs.

In Q14, it is not dc power that cable companies use to power the trunk and line extender amplifiers. The power source is a 60 Vac transformer directly inserted on the line through a power inserter type of splitter. This 60 volts might drop, due to line resistance, to as little as 30 volts, which most line equipment will still work with. The answer to Q14 is b.

The term used to identify interference which is of the same frequency as that selected (channel 13 from afar interfering with a local channel 13, for instance) is *co-channel*. Adjacent channel interference is usually the channel immediately above or immediately below the desired channel. If an adjacent channel is much greater in size it will override the desired channel, usually causing trash across a video picture. In radio it can cause break up, or the desired and the interfering channel periodically taking control and breaking in on each other. Intermodulation interference is best known for its effect on high-fidelity sound equipment where the circuitry, speaker or other components are unable to exactly reproduce the wanted sound patterns. The resultant harmonic or mechanical misrepresentation is objectionable to the ear, thus equipment makers constantly attempt to reduce it (Fig. 17-5).

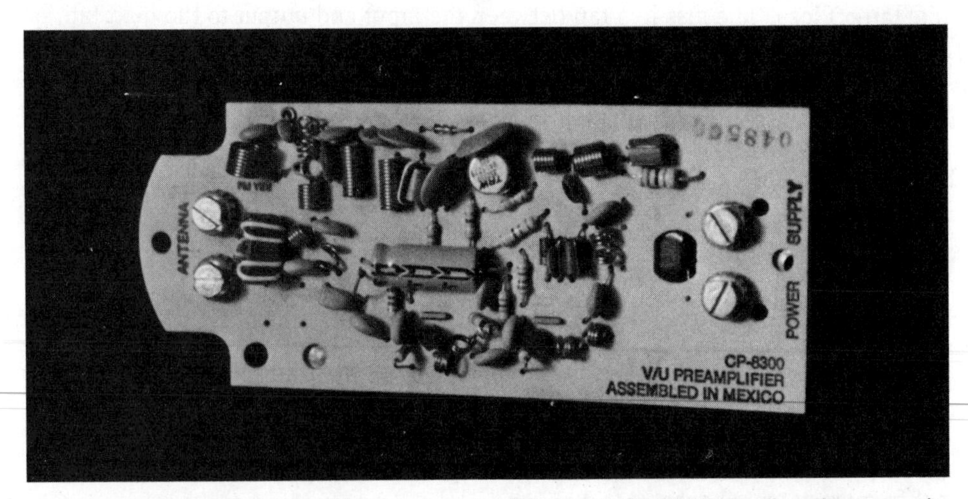

17-5 Mast-mounted preamplifier section of a Winegard CP 8300 showing three transistors, filter capacitor, and other RF components. This version requires twin-lead transmission cable from the antenna terminals as well as twin lead out and down to the in-house 14 V power supply.

In Q16, most antenna preamps in the past have used a 14 – 18 volt ac power supply attached directly to the down-line from the mast mounted pre-amp. The ac component that the TV signals are riding down on is removed with a high-pass filter, located in the power supply section of the pre-amp. This 14 – 18 Vac is used in the pre-amplifier as the power source, being rectified and filtered prior to being used by the transistors. The mast mounted amplifiers contain a power supply just like other electronic products. That power supply will be a diode rectifier to cut off half of the 60 Hz power supply waveform, then filter circuitry containing a large value smoothing-capacitor. Lightning is the most likely cause of the diode shorting or opening, as well as destroying the amplifier transistors. Infrequently, the capacitor opens which gives the symptom of 60 Hz bands in the video and buzz in the sound on all TV channels (Fig. 17-6).

17-6 The power supply section for a mast-mounted TV preamplifier. The top F81 coax connector has a small value capacitor feeding RF to the bottom connector which prevents the power frequency (60 Hz) from passing, but allows RF TV signals to be passed. The 120 V step-down transformer produces the 14 Vac to power the mast-mounted preamplifier. Note the preamplifier in Fig. 17-5 will not work with this power supply, as this one requires RG-6 or RG-59 transmission cable.

Some mast-mounted amplifiers are now manufactured that do not use 14 – 18 Vac. Instead they rectify the 14 – 18 Vac in the power supply section (which is normally mounted inside the home). Then power of 12 – 24 Vdc is supplied up the coax cable between the power supply and amplifier section mounted near the actual antenna elements. Eliminating the power supply components from the mast-mounted portion of the pre-amp reduces the percentage of service incidents requiring tower or roof work. 10 dB and 20 dB in-line amplifiers are also commonly used in satellite systems. Because the TVRO receiver already is providing LNB power from the house to the dish outside and this power is 18 Vdc usually, a small in-line amplifer can be inserted directly in the line, usually out at the dish, to increase the LNB's output IF signal. This can overcome the extreme line losses experienced when 300 to 400 foot RG-6 lengths must be used.

Matching the video distribution components in Q17:

a = 4
b = 2
c = 3
d = 5
e = 6
f = 1

As a brief explanation, b is the value of a 4-way signal splitter: – 7 dB. The common amplifier gain value is 24 dB. The compensation attained by correcting long-line losses with a line compensator is 8 dB on low VHF and 0 dB on high VHF.

A bandsplitter is used to separate UHF and VHF signals, most commonly needed at the inputs to VCRs and TV sets, but also used in SMATV and cable head-ends to direct each band of channels to where it can be processed and to keep others out. To reduce signals, as you must do when combining master antenna signals, cable signals, etc., so that all the channels are very near the same voltage level, an attenuator is used. These come in sizes such as 3 dB, 6 dB, 10 dB, 16 dB, and 20 dB.

In Q18 you should have some idea of the signal level to be expected out of a VCR, disc player, satellite receiver, or any other product intended to drive a TV set. While the TV is designed to provide a snow-free picture with at least 1000 microvolts of antenna terminal input signal voltage (0 dBmV), most VCRs and the like have enough output to drive two or three TVs without having snowy pictures. Common outputs are 6 – 10 dB. The best answer to Q18 is c.

The difference between the old copper-braid RG-59, and foam-insulation modern RG-6 coaxial cable is significant. At UHF frequencies, RG-59 copper-shield cable will have 50 percent more loss. In TVRO work, the initial dish installation in the early 80s required only RG-59, as the signal from the LNA/downconverter was a low 70 MHz band no matter what frequency or channel you were on. Modern TVRO receivers send from the LNB (low-noise block downconverter) a 500-MHz wide band containing all 24 (or 36 in Ku band) channels. This band is much higher than 70 MHz. It is 950 MHz to 1450 MHz, or over 10 times as high of a frequency band. Attempting to use the old RG-59 cables when upgrading the system, results in inadequate IF signals in most cases and the RG-59 must be changed.

All technicians who attempt to do antenna work should have access to a field strength or signal-level meter. There is no way to perform MATV or cable work without one. Prices for one run from a low $500 to several thousand dollars. Most FSMs have an analog meter, which at full-scale can read 100 microvolts or can be calculated to read 1000 microvolts. If the design is for antenna work where you might well be interested in signals of 10 microvolts or even smaller (for instance in checking for cable leaks), you will most likely need a 100 microvolt meter. For cable work, a 1000 microvolt full-scale reading is satisfactory. To read higher levels of signal, 20 dB attenuator switches are provided, or a combination of 20 and 10 dB switches. Without external attenuators, most FSMs will check −40 to +40 dBmV signals (Fig. 17-7).

17-7 Channel Master's Video Control Center (front and back views) with multiple signal sources and multiple viewing areas in most homes, these devices are very useful to installation technicians.

Video distribution technicians need to know all about antennas used for TV and radio. They need to know what each component is used for and its characteristics. They need to know how to hook any product up to today's modern TVs, VCRs, and high-tech A/V equipment. They need to know how to distribute the signals to a second viewing area, or to an apartment complex. In doing so, knowledge sufficient to service larger cable systems will be gained. While many techs in consumer electronics have never been introduced to satellite reception equipment, they should learn about it anyway. It is a part of your cable company, your local radio station, business and entertainment establishments.

18 Radio communications option

RADIO COMMUNICATIONS AND *TELECOMMUNICATIONS* ARE TWO TERMS that frequently are used interchangeably. The CET exam committee and the Electronics Technicians Association divide the two topics as follows:

Radio communications Includes two-way amateur radio, business radio, ham radio, maritime and aviation navigation and communications equipment, walkie-talkies and short-range radio, military communications radio, and radar.

Telecommunications Includes television, any distribution of TV and telephone signals, cellular phones and pagers, data communications, microwave, and fiberoptic communications.

Obviously there is a lot of overlap in the usage of the two terms and in the equipment that technicians maintain. For instance, a cellular phone system uses FM transmitters and receivers (Radio Communications), but it also performs telephone communications (Telecom).

Where does radar, as used by airports, weather services, and avionics fit in? Radar uses video, but it isn't television or telephone. Often radar is taught in telecommunications courses as well as in avionics and military radio communications classes. Take your pick.

Quiz

Q1. Match the following, as used in RadComm:

a. CAD — 1. phase-locked loop
b. ARRL — 2. Society of Broadcast Engineers

c. SBE
d. PLL
e. SINAD

3. American Radio Relay League
4. computer assisted dispatching
5. Communications Assn. Dealers
6. Signal/noise/distortion compared with noise/distortion ratio
7. single integrated network to attenuate distortion

Q2. The amount of RF signal required to reduce noise at a radio's input by 20 dB is:

a. 20-dB sensitivity.
b. 20-dB quieting.
c. 20-dB signal to noise.

Q3. In business radio, AVL stands for:

a. automatic volume level.
b. automatic vehicular location.
c. attenuated volume level.
d. American Low Frequency Vehicle Low Power Assn.

Q4. A base station remote can be connected to a telephone line. When this is done the output of the base station should be:

a. 0 dBm at 75 ohms.
c. 18 dBm at 600 ohms.
b. 18 dBm at 300 ohms.
d. 6 dBm at 300 ohms.

Q5. Transmission cable used for CB and other communications transceivers is the same impedance cable as that used for TV—72 ohms.

a. True
b. False

Q6. In Fig. 18-1, if the potentiometer wiper moves towards the + voltage end, it would:

a. cause the squelch circuit to prevent stronger signals from turning on the audio.
b. cause the squelch circuit to prevent weaker signals from turning on the audio.
c. have no effect on the squelch because that potentiometer is the volume control.

Q7. With a stronger signal being received, the AVC voltage in Fig. 18-1 would be:

a. more positive.
b. more negative.

Q8. To calculate the efficiency of a transmitter output stage, one could compare the power being transmitted via the antenna (for instance 4 watts for a properly operating CB radio), with the input signal to that stage (for instance a 1-volt RMS signal across the base circuit load resistor).

a. True
b. False

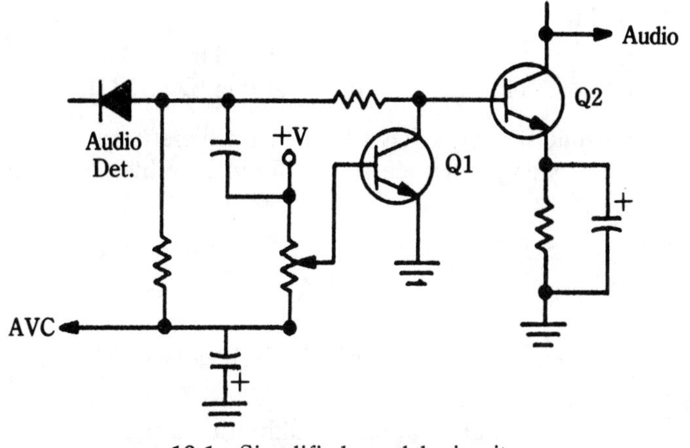

18-1 Simplified squelch circuit.

Q9. Suppose the transmitter transistor is drawing 5 amps through the emitter resistor (0.47 ohm). How much power is the output stage consuming?

a. about 12 watts
b. about 2.5 watts
c. about 60 watts

Q10. If the transmitter output stage produced 20 watts of power at the antenna and the output stage was absorbing 40 watts of power from the power supply, what might the efficiency of the stage be?

a. Unable to determine.
b. 25 percent
c. 50 percent
d. 100 percent

Q11. The RF stage of most radio receivers normally has higher gain than the IF stage(s).

a. True
b. False

Q12. The RF stage of most radio receivers normally has wider bandwidth than the IF stage(s).

a. True
b. False

Q13. SCPC might refer to satellite transponder stations that are not subcarriers of a separate video channel carrier.

a. True
b. False

Q14. A klystron is used primarily in:

a. satellite receivers.
b. satellite transmitters.

Q15. In the 800-MHz business radio band, the transmit and receive frequencies allowing full duplex are:

a. The same.
b. separated by 455 kHz.
c. separated by 10.7 MHz.
d. separated by 45 MHz.

Q16. Ionized atmospheric layers have been used beneficially in broadcasting. They also have been the source of interference. Which layer reflects UHF frequencies?

a. E layer
b. R layer
c. Z layer

Q17. When used in communications work WWV means:

a. the abbreviation for the upcoming world war five.
b. NBS broadcasts suitable for use in calibrating frequency meters in shops.
c. NPR's Washington, D.C. origination station.
d. The NBC affiliate in Charleston, West Virginia.

Q18. Crystals, as used in oscillators in radio circuits, can be made to oscillate if:

a. they are in parallel with a suitable capacitor.
b. they are in series with a suitable capacitor.
c. they are in series or parallel with a suitable capacitor.

Q19. The term *parity* is:

a. a type of computer interface.
b. a bit at the end of a binary word used for error checking.
c. the locking pulse used in simulcast systems.
d. equal power levels at both transmitter locations in two-way broadcasting.

Q20. Metal oxide varistors are most often found

a. in PLL cirucits.
b. in VCO circuits.
c. in power supply circuits.
d. in surge protector circuits.

Quiz explanation

CAD means computer assisted dispatching. ARRL is the abbreviation of the American Radio Relay League, a nonprofit association to which most amateur radio buffs belong. The SBE is the Society of Broadcast Engineers, an association for those who work at radio and TV stations as technicians and engineers.

PLL stands for Phase-Locked Loop, today's circuitry that has virtually eliminated drifting of radio frequencies. SINAD is a term used in two-way radio that is similar to the standard signal to noise ratio. SINAD gives a signal/noise/distortion ratio as compared with the noise/distortion ratio.

The term used to indicate the amount of RF signal required to reduce noise at the radio's input by 20 dB is 20-dB quieting. Thus the answer to Q2 is b.

AVL stands for automatic vehicular location; the answer to Q3 is b.

When connecting to a telephone line, the base station should have an output impedance of 600 ohms and a signal level of 18 dBm.

While some communications coaxial cable appears similar to the RG-59 and RG-6 used in cable/MATV/TV, which has 72-ohm impedance, most communications radio coax is 50-ohm impedance cable (RG-58 for instance). The answer to Q5 is b.

The squelch circuit in Fig. 18-1 operates by causing the Q1 transistor to turn on at some point, which shorts the base of Q2 to ground effectively, thus biasing Q2 to cutoff. This shuts off the audio. Any weak signals picked up by the antenna and RF/IF sections will be shorted to ground at the Q2 base.

Should some out-of-range communications still be heard, the operator can adjust the potentiometer toward the V+ end, which will bias Q1 to be turned on when even stronger signals are received and the AVC dc level goes further negative. When very large in-range local signals are received, the AVC goes much more negative, but the positive portion of the rectified IF envelope will be sufficient to overcome the small collector voltage on Q1, turning off Q1 and allowing Q2 to amplify the audio signal. Answers a and b for Q6 are both right, depending on how you interpret them. The CET exam doesn't contain such difficult-to-explain answers. The squelch is changed when the arrow is moved up to V+ so that signals of an even higher amplitude will be rejected. Answer a is best.

Question 7 requires you to decide what happens with the AVC voltage in Fig. 18-1 when stronger signals are received. As stronger signals are received the audio detector, acting just like a power supply rectifier, draws more current in a direction opposite the diode symbol arrow, thus pushing more electrons down into the AVC circuit storage capacitor, which is an electrolytic installed with positive to ground.

The efficiency of a transmitter is calculated by determining the output power that stage is producing, using a watt meter, then calculating the power consumed by the stage. Because power output stages (tube or transistor) ordinarily use large amounts of power as opposed to signal voltage amplifiers, efficiency is important. To calculate the power consumed, it is common to measure the emitter or collector current, then multiply the current by the voltage drop across the circuit. If 1 amp is flowing through the transistor from a 12-volt supply, then ($P = EI$) $12 \times 1 = 12$ watts input power. If the signal output is measured at 6 watts, then the efficiency is 50 percent. The input signal to a power output stage is important in that most transmitters operate class C, meaning that it is the signal excursions that overcome the preset transistor (or tube) bias, causing current to flow. Without input signal, the stage draws no current at all. However, the input signal has no bearing on the stage efficiency, so Q8 is false.

In Q9, the 5 amps is dropped across the transistor stage. You can calculate the wattage dissipated by the emitter resistor ($P = IR$), or 12 watts (answer a) and the entire stage ($P = EI$) or 60 watts.

Question 10 is the same question asked another way. Answer c is the correct response.

The RF stage of a radio receiver must be able to receive signals over a broad range of frequencies. In AM broadcast radio this is over 1 MHz of bandwidth, from 535 kHz to 1605 kHz. Gain is inversely proportional to bandwidth. The higher the gain, the smaller the bandwidth. Because the RF stage must be able to respond to signals over a wide range of frequencies, it has much lower gain than the IF stages, which in AM radio have only to respond to a narrow band of frequencies, perhaps 10 kHz. That is one of the most important differences between the old TRF receivers, which merely had multiple stages of RF amplification before the detector, and the superheterodyne receiver that converts all RF signals to a common IF frequency. The correct answer to Q11 is b.

The answer to Q12 is a. Most RF stages do have broader bandwidth than the IF stages.

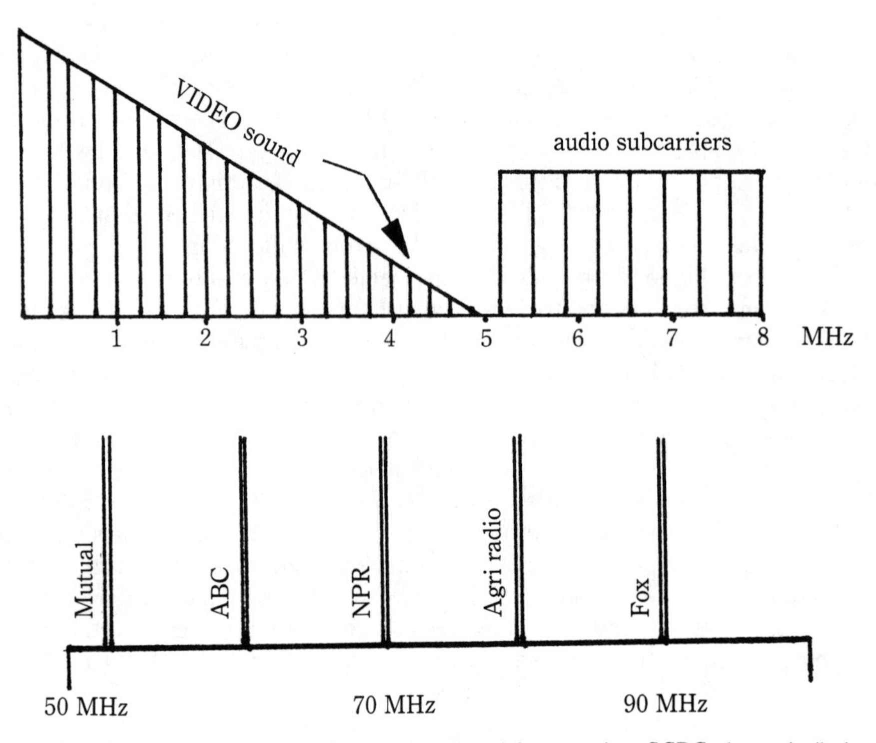

18-2 Video channel subcarriers (above) at baseband frequencies. SCPC channels (below) at IF frequencies.

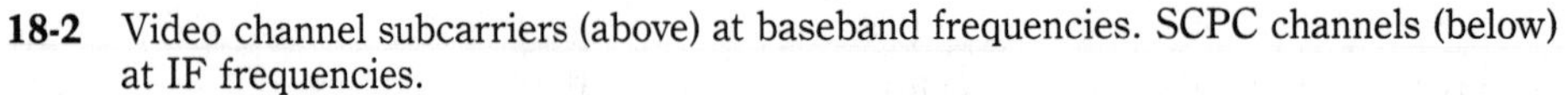

Question 13 asks if you know what the term SCPC means. It is Single Channel Per Carrier, answer a. Today many communications channels are dependent on some carrier which has no relationship with the voice of video message of that carrier. Some satellite communications channels carry only a single video service with the accompanying sound. It might be a station like ESPN video and audio. The audio is actually a subcarrier that would not exist without the video carrier frequency. There are other video channels on satellite that carry the video channel's voice subcarrier, and also one or several radio station transmissions. These radio stations are subcarriers, dependent on the carrier of the video transmission. There might be a half-dozen stereo subcarriers sharing spectrum with a single video signal. The allotted spectrum for a single channel is 36 MHz (with a 2-MHz guard band on either side, for 40 MHz total). The video and its accompanying audio might use 24 MHz of the 36-MHz allowable band, and each of the subcarriers might use only 60 kHz of space. So several radio voice channels, each with full 20 Hz – 20 kHz frequency response range, can be used on a single video channel slot.

These are not SCPC signals; they are subcarrier signals. SCPC signals are similar to ordinary FM broadcast signals. They do not rely on another carrier, but are independent stations occupying about 60 kHz of spectrum each. Therefore, one satellite channel 36 MHz wide can support several dozen radio services such as Mutual, the Bluegrass Network, Indiana Agri-Radio, and so forth. The answer to Q13 is a.

A *klystron* is a high-power device used in radar and other transmitters. It is a vacuum tube in design, using resonant cavities and mechanical design to accomplish its transmitter purpose. The answer to Q14 is b.

There are some things that just have to be memorized. One is the choice of

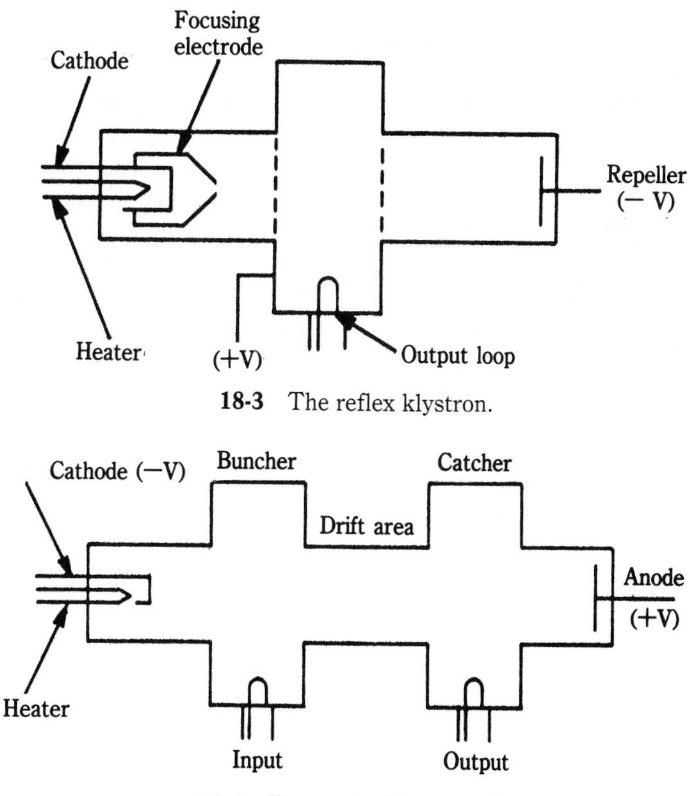

18-3 The reflex klystron.

18-4 Two-cavity klystron tube.

receive and transmit frequencies in the 800 MHz two-way mobile phone band. The separation is 45 MHz; answer d to Q15.

As radio broadcasting evolved, the phenomenon that caused the transmitted signals to bounce off of the several ionized atmospheric layers was recognized and put to use. "Skip" allowed broadcast frequencies to travel in space until they encountered one of the definable layers that act somewhat like energized neon, or like the Aurora Borealis. The broadcast signals bounce off the layer at an angle and are then redirected back to earth, often at a considerable distance farther away than is possible with the terrestrial direct transmission. The layers have been given letter designations. They can be at different altitudes depending on whether it is night or day. Each affects RF signals differently, depending on the signal frequency. The layers affecting TV signals are mainly the F and E layers. Answer a is correct for Q16.

Question 17 asks if you recognize the call letters for the U.S. National Bureau of Standards: WWV. Exact time reference signals are broadcast as a public service on several frequencies that can be used by anyone. WWVH is the identical service broadcast from Hawaii.

Capacitors are used to complete the crystal oscillator circuit and to maintain the frequency at a precise point. The crystal oscillates once a voltage is impressed on one side of it. The capacitor may be in series or in parallel. The answer to Q18 is c.

Parity, in Q19, is the term used for the bit at the end of a binary word used to check for errors.

MOVs, in Q20, are the metal oxide varistors used in surge protectors. MOVs are high impedance until a certain predetermined voltage is impressed across them. At that point the MOV becomes a low impedance, shorting the incoming voltage to ground. MOVs can range from a breakover voltage of around five volts to over 3000 volts. The capabilities so far as amperage shunting are also over a wide range, with 25 amps to over 70,000 amps peak possible. Obviously the extreme amperage capabilities can be for pulse durations in the microseconds. Reaction time for MOVs is a specification in the nanosecond range. The correct answer to Q20 is d.

19 The Avionics endorsement exam

THE AVIONICS SECTION OF THE CET EXAM PROGRAM IS NOT AN "OPTION" choice. Rather, it is an endorsement to be added to either the Radio Communications option or the Telecommunications option. Avionics is a specialized area of electronic service. The CET committee of ETA-I has as its intention the certification of technicians in communications radio, but to not require all radio-communications and telecommunications technicians to be totally knowledgeable in avionics. On the other hand, those working for aircraft communications companies, airports, government agencies (like the FAA and in FAA certified electronics service shops) need to have the specific knowledge of the specialized equipment used in avionics.

Quiz

Q1. VOR stands for:

a. velocity operational radio.
b. VHF overland radar.
c. variable oscilloscope reading.
d. VHF omni range.

Q2. The VHF band used for aviation communications is:

a. 30 MHz to 300 MHz.
b. 300 MHz to 3000 MHz.
c. 118 MHz to 135.95 MHz.
d. 118 MHz to 125.95 MHz.

Q3. A VOR site transmits a beam signal from one antenna and an omnidirectional signal from a second antenna.

a. True
b. False

Q4. A radar pulse will travel out one nautical mile, be reflected, and return in what length of time?

a. 6.2 microseconds
b. 12.4 microseconds
c. 50 microseconds

Q5. DME stands for:

a. data metric entries.
b. distance measuring electronics.
c. distance measuring equipment.
d. direction measuring equipment.

Q6. Loop antennas are typically:

a. nondirectional.
b. omnidirectional.
c. unidirectional.
d. bidirectional.

Q7. ILS systems consist of three parts: localizer, glide slope, and:

a. resolver amplification.
b. goniometer.
c. marker beacons.
d. gyros.

Q8. What two modulation frequencies are used in instrument landing system transmitters?

a. 60 Hz and 120 Hz
b. 90 Hz and 150 Hz
c. 108 MHz and 111.8 MHz
d. ILS systems do not use modulation.

Q9. ILS system localizers provide what type of information?

a. vertical position
b. horizontal position
c. range information
d. azimuth information

Q10. Glide-slope ground equipment uses two transmitters in the 329-MHz to 335-MHz range, modulated by what two frequencies?

a. 60 Hz and 120 Hz
b. 90 Hz and 150 Hz
c. 108 MHz and 111.8 MHz
d. 400 Hz and 1200 Hz

Q11. Marker beacons are broadcast by the:

a. aircraft.
b. airport.
c. regional FAA center.

Q12. A marker beacon receiver provides violet, amber, and white indicators to show distance from the runway. It also provides three different audible tones.

a. True
b. False

Q13. MLS is a newer system for landing communications that use TRSB (time referenced scanning frequencies) and operates in the C-Band.

a. True
b. False

Q14. Radar systems use pulse coded spacing to:

a. select the desired aircraft transponder.
b. reject the undesired aircraft transponder.
c. allow all local area returns to be received.

Q15. If altimeters need a different binary code to represent the barometric pressures from −1000 feet to 127,000 feet, in 100-foot increments, how many codes would be needed?

a. 128,000
b. 12,800
c. 1280
d. 1024

Q16. Weather radar with a transmitter output of 10 kW uses a __________________ to produce the power.

a. gyro
b. magnetron
c. MLS
d. Shockley diode

Q17. Match the following frequencies used in avionics:

a. 9375 MHz	1. MLS
b. 1030 MHz	2. glide-slope transmitter
c. 5090 MHz	3. radar interrogator pulse
d. 400 – 1300 – 3000 Hz	4. weather radar
e. 329 MHz to 335 MHz	5. marker transmitter
f. 108 MHz	6. localizer frequency

Quiz explanation

VOR stands for VHF Omni Range. It is a method of electronically determining the location of an aircraft relative to a land location, usually an airport.

It uses two transmitter antennas, one fixed and one rotating. The phase relationship between the rotating "beam" of signal from one antenna and a second fixed antenna's omnidirectional signal allows the receiver circuitry in the aircraft to display the correct bearing. The answer to Q1 is d. The frequency band used for VOR stations is immediately below the 180 channel VHF aircraft band, about 108 MHz to 119 MHz (Table 19-1).

Table 19-1. A few DME channels paired with VOR frequencies.

Channel No.	Frequency of Paired VOR Station in MHz	DME Interrogation Frequency in MHz	DME Reply in MHz
17X	108.00	1041	978
17Y	108.05	1041	1104
18X	108.10	1042	979
18Y	108.15	1042	1105
19X	108.20	1043	980
19Y	108.25	1043	1106
—	—	—	—
125X	117.80	1149	1086
125Y	117.85	1149	1212
126X	117.90	1150	1087
126Y	117.95	1150	1213

In Q2 the VHF aviation band is 180 channels of 50-kHz steps within the 118 to 135.95 VHF band. Answer a is correct in that the VHF aircraft band is between 30 MHz and 300 MHz. That is the VHF band in its entirety. VHF low-band TV is 54 – 88 MHz. The TV high-band is 174 – 216 MHz. So the spectrum for aircraft VHF communications is above the FM broadcast band (88 – 108 MHz), above VHF low and below VHF high-band. The band used is almost 18 MHz wide. Answer c is the best response to Q2.

In Q3, the beam signal is rotated 360 degrees at 1800 rpm. The second antenna is fixed. In the receiver the phase difference between the two antennas is detected. An indicator shows whether the aircraft is to the left or the right of the desired bearing. Another indicator shows whether the aircraft is approaching or leaving the VOR station. The correct response to Q3 is a (Fig. 19-1).

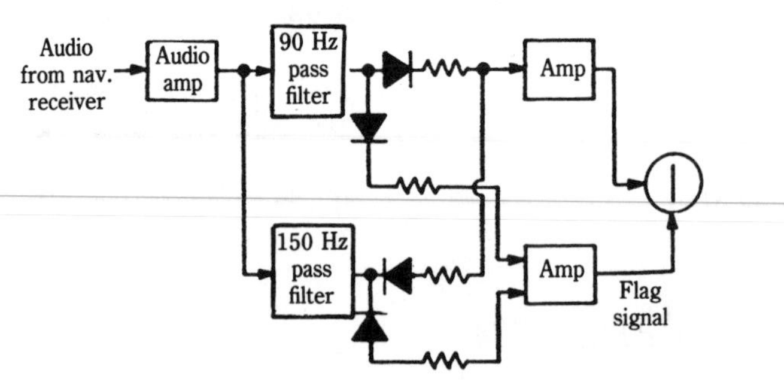

19-1 Localizer block diagram.

One thing that remains constant for a technician is the fact that a radar mile (the time it takes for an electromagnetic pulse to travel one mile and return) is 12.4 microseconds, answer b to Q4. It follows that it must take an electronic signal 6.2 microseconds to travel one nautical mile then.

In Q5, DME stands for Distance Measuring Equipment. In Q1, Q2, and Q3 the direction of the aircraft relative to the VOR (airport) site was important. DME determines the distance from the site, or the distance from a waypoint tangent to the VOR site. DME uses the UHF band not the VHF, as was used in the VOR direction finding equipment (Fig. 19-2).

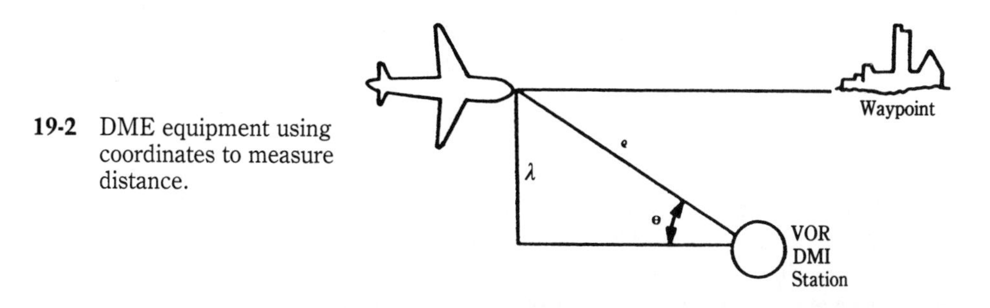

19-2 DME equipment using coordinates to measure distance.

DME ground equipment transmits an interrogation frequency which then results in a reply from the aircraft distance measuring equipment. The ground equipment then returns pulses to the aircraft DME that calculates the distance from the aircraft to the next waypoint. Frequencies used for this operation are near 1 gigahertz, still UHF, but above the TV UHF band (channel 83 is 890 MHz). The interrogator and reply frequencies are always separated by 63 MHz.

In Q6 loop antennas have always been the heart of direction-finding equipment. In their basic form, they are bidirectional. This is similar to a half-wave dipole, which has a figure 8 polar pattern. Loop antennas for low- and high-frequency bands are usually coils of wire. These also have a figure 8 pattern if the coil is in the shape of a square or rectangle. The answer to Q6 is d.

By adding the signal from a fixed antenna and a loop together in certain ways, the loop can be made unidirectional. A more exotic configuration, but an outgrowth of a basic loop, is a goniometer (Fig. 19-3).

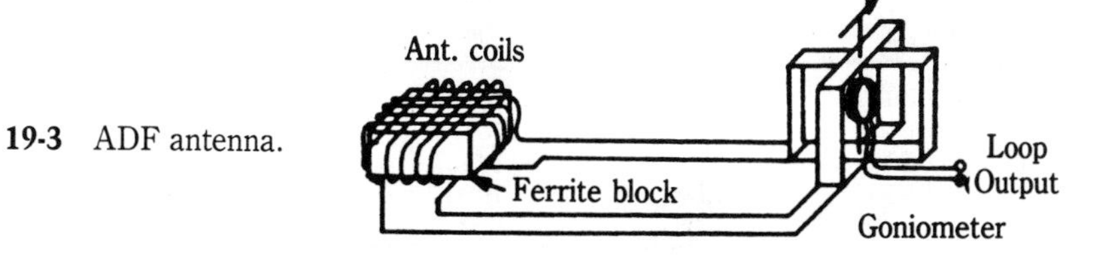

19-3 ADF antenna.

The term ILS in Q7 stands for Instrument Landing System. You are asked if you know the third major component or function of the system. The localizer portion tells you the horizontal direction the plane might have at a given moment, relative to the runway. The glide slope gives you the vertical position or altitude as the aircraft proceeds down the desired path to the runway touchdown. The third part is the marker beacons that give range information, or the distance from the runway: Answer c (Fig. 19-4).

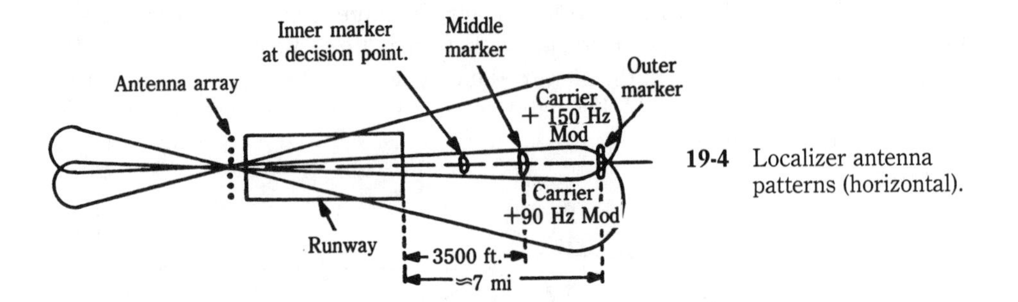

19-4 Localizer antenna patterns (horizontal).

In Q8 you need to know that ILS ground equipment transmits two teardrop-shaped overlapping beams up the runway towards the approaching aircraft. By modulating these two beams (one with 90 Hz and one with 150 Hz), the aircraft's receiver can detect which side of the runway the plane has veered off to, or whether it is in the overlapping portion of the beams, which is the center of the runway. The receiver gives this indication, allowing the pilot to make the proper correction in direction of the aircraft.

In Q9 you must know that the ILS system localizer provides horizontal position information for the pilot. See Table 19-2.

Table 19-2. Localizer/glide-slope pairs.

Frequency in MHz	Type of Service	DME frequency in MHz	Glide-slope frequency in MHz
108.00	VOR	1041	
108.05	VOR	1041	
108.10	Localizer	1042	334.70
108.15	Localizer	1042	334.55
108.20	VOR	1043	
108.25	VOR	1043	
108.30	Localizer	1044	334.10
108.35	Localizer	1044	339.95
—	—	—	—
111.80	VOR	1079	
111.85	VOR	1079	
111.90	Localizer	1080	331.10
111.95	Localizer	1080	330.95

In Q10, the glide-slope equipment operates much like the localizer with two transmitters, one modulated with 90 Hz and one with 150 Hz. The two beams overlap, but only for a small portion of the beams—2.5 to 3.0 degrees. When the received signals come from both the 90- and 150-Hz modulated transmission beams, the equipment recognizes the aircraft is exactly within the glide-slope beam channel. Actually, the equipment gives an audible tone to indicate that the aircraft is too high or too low. When in the correct vertical position, at the narrow path where the two transmissions overlap, there is a null produced and the pilot recognizes that he/she is on the correct path.

The localizer transmissions provide horizontal beams that overlap for a very narrow portion of the teardrop-shaped beams, letting the pilot know he is aimed at the center of the runway. The glide slope works the same way, except the beams are vertically one on top of the other with the overlapping portion aimed up the glide-slope optimum path (Fig. 19-5).

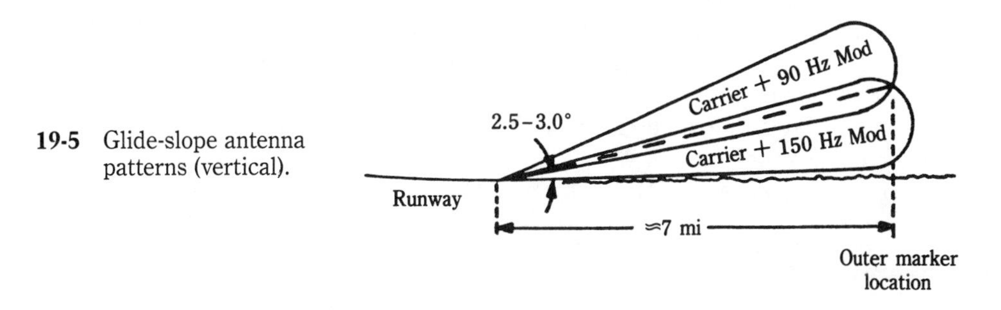

19-5 Glide-slope antenna patterns (vertical).

In Q11, marker beacons are discussed. They are transmitted by the airport. The outer marker produces a 75-MHz fan-shaped beam aimed straight up. An approaching aircraft flying over this outer marker will receive that signal, but not the other two marker beacons. The outer marker is modulated at 400 MHz and this is received and heard by the pilot. In addition to the tone, a blue-violet lamp indicator turns on. As the plane leaves the outer marker and approaches the middle marker, an amber indicator lights up and a faster audio code is heard (1300 dot/dash frequency). The inner (and last) marker lights a white indicator and produces an even faster dot/dash audio tone (3000 Hz). The answer to Q11 is b; and Q12 is a (Fig. 19-6).

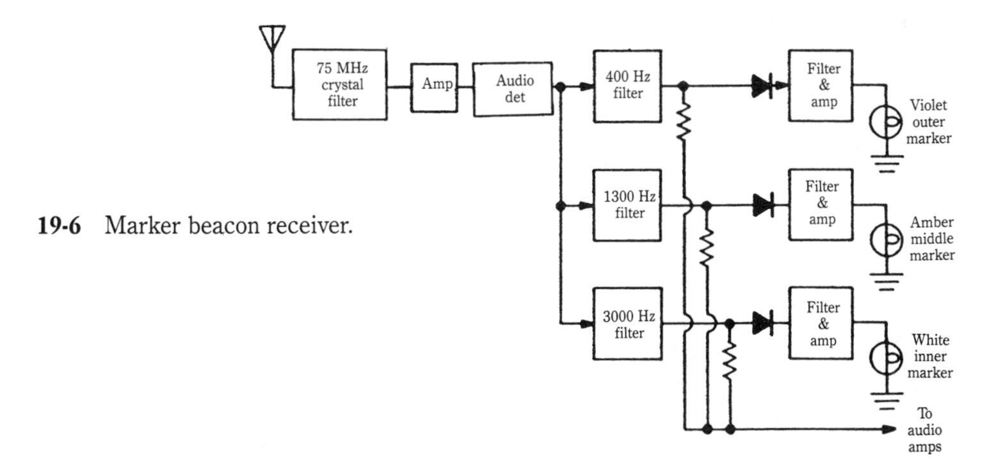

19-6 Marker beacon receiver.

Question 13 is true. The time-referenced scanning beam (TRSB) operates in the C-band (200 channels using a band from 5031 MHz to 5090 MHz). This system uses radar techniques to precisely determine altitude, asimuth, and range. It can be used for automated auto pilot systems or to simply drive instrument panel indicators.

In Q14, radar systems use pulse-coded-spacing to differentiate between other returns and the desired aircraft. The spacing identifies the aircraft so that the radar operator knows exactly which return blip on the screen is the one he is in communication with. An SPI (Special Position Identifier) pulse can also be sent by the pilot to highlight the desired indication on the screen.

In Q15, 1280 is 128,000 divided by 100. Altimeters consider −1000 feet the lowest altitude they will ever need to measure, and 127,000 the highest. The difference is 128,000. If you need the 1280 increments and are using a binary system, you need 1280 different indicators; 1024 is not enough. All the information you can get from a 10-data-line setup is 1024 (2 to the 10th power is 1024). Therefore c is the correct answer to Q15.

You need to know for Q16 that a magnetron has been used for many years (at least 50) to produce the high power (10 kW or more) used by aircraft weather radar. That radar operates at 9375 MHz (9.275 gigahertz), which was chosen because it is excellent for reflecting off precipitation.

The matches in Q17 are as follows:

a = 4
b = 3
c = 1
d = 5
e = 2
f = 6

20 The Satellite TV option

THE TVRO, OR SATELLITE TV OPTION, IS FOR CET CANDIDATES WHO ARE technicians working primarily with satellite communications equipment (Fig. 20-1). They should be aware of the technology used to receive satellite signals, the hardware operation, repair procedures, and interfacing techniques. The Electronics Technicians Assoc., Int'l., also administers a certification examination for satellite equipment installers called the CSI, or Certified Satellite Installer examination. The CSI is expected to be an expert in installation and troubleshooting, but he/she is not required to be a board-level component bench repair technician. Satellite communications are a part of practically all electronics fields. Computers, telecommunications, consumer electronics, cable, and other areas are in many ways now using satellites. Technicians will do well to have at least a minimal understanding of the technology used.

Quiz

Q1. The Clarke Belt is a practical location for geostationary or geosynchronous satellites, but other orbits, more distant or nearer to earth, could be used for these seemingly stationary satellite positions. This would be done by speeding up more distant orbiting satellites, or slowing down nearer positioned ones.

a. True
b. False

Q2. The bands for frequencies used for C band satellite communications and for Ku band are both 500 MHz wide.

a. True
b. False

20-1 A 10-foot diameter black mesh polar-mounted satellite dish with linear motor drive.

Q3. The amount of signal loss from the orbiting satellites to the U.S. land is approximately:

a. 50 dB c. 196 dB
b. 92 dB d. 296 dB

Q4. Because of the differences in dish reflector gain, as well as transponder power output and cable wire losses from the reflector to the receiver, the TVRO receiver gain ranges from:

a. 6 to 36 dB c. 36 to 96 dB
b. 20 to 50 dB d. 50 to 96 dB

Q5. Dithering (in TVRO) is a process used to:

a. jog the dish position exactly onto the desired satellite boresight.
b. reduce the effect of noise in the video signal.
c. move the feedhorn antenna rotor to the precise angle.
d. automatically center the video fine tuning on TVRO channels.

Q6. Match the following:

a. 36 Vdc	1. coax switch actuating voltage
b. 18 Vdc	2. video signal level output
c. 5 Vdc	3. receiver circuitry power supply
d. 5 volts digital pulses	4. servo motor positioning energy
e. 2000 microvolts	5. LNA power
f. +12 Vdc, −12 Vdc	6. dish drive motor voltage

Q7. Detector circuits in the TVRO receiver will be what type?

a. FM
b. AM

Q8. Composite output from the receiver contains video and sync information but not audio.

a. True
b. False

Q9. Filtered video output contains audio subcarriers that have been de-emphasized.

a. True
b. False

Q10. Background flickering of the brightness level might indicate a defect in the ______________________ circuitry.

a. dedithering
b. modulator
c. RF section
d. LNB or LNA

Q11. Fuses found in satellite receiver/positioners might be expected to be of what size?

a. 2 amp motor – 1 amp receiver power supply – 1/2 amp LNB
b. 2 amp receiver power – 1 amp motor – 1 amp LNB
c. 2 amp motor – 5 amp receiver – 1 amp LNB

Q12. SCPC signals and audio subcarriers may both be found on a single C band channel.

a. True
b. False

Q13. In aiming a polar mount satellite dish, the polar bar is pointed directly at the North Star.

a. True
b. False

Q14. Once the polar mounted satellite dish polar bar is set to the correct latitude angle, the declination adjustment should be set. This causes the reflector to be then aimed:

a. higher
b. lower

Q15. An external filter that clips off the low and high ends of the pass band of the IF signal is called:

a. a notch filter.
b. a band-pass filter.
c. a high Q trap.
d. a resonant cavity microwave trap.

Q16. Ku band transponders can have 16 or 32 channels rather than the 24 now found on C band.

a. True
b. False

Q17. A stand-alone consumer videocipher II decoder allows adjustment of the ____________________ audio output level.

a. descrambled
b. unscrambled

Q18. 18 Vdc is measured on the coaxial cable center conductor after it is disconnected from the LNB (Fig. 20-2). The voltage drops to less than 5 volts when the LNB is again connected to the coax. Which of the following is not the cause?

a. high resistance in the LNB cable.
b. defect in the receiver power supply.
c. shorted regulator in the LNB.
d. open regulator in the LNB.

20-2 Entrance waveguide showing Low Noise Block downconverter cylindrical pick-up element inside. RF signals from the orbiting satellites are reflected off the dish into this LNB waveguide and onto the small pedestal pick-up. The pick-up is then connected to the input of the LNB's first RF stage of amplification.

Q19. After you connect a rooftop antenna system to a satellite receiver, UHF TV stations become extremely weak. What is wrong?

a. The modulator has a defect.
b. The modulator is not designed to pass UHF frequencies.

Q20. The f/d ratio of a parabolic TVRO dish reflector is:

a. the focal distance compared to the dish depth.
b. the focal distance compared to the dish diameter.

Quiz explanation

Question 1 is false. More distant orbits would require a slower speed to remain in orbit. Then the satellite constantly would appear to be moving west, rather than appearing permanently over the same spot on the equator. The opposite holds true for an orbit nearer the earth than 22,347 miles.

Question 2 is correct. While the frequencies used are different at 3.7 to 4.2 gigahertz for C band and 11.7 to 12.2 for Ku band, the width of the frequency spectrum is 500 MHz in both cases.

You must understand just how tiny the signals from the satellites are for Q3. The loss, due to dispersal of the transmitted signal, along with a small amount of atmospheric attenuation, is approximately 196 dB. To compensate for that loss, a dish reflector might have a gain of 40 dB, and LNB has 60 dB and the receiver from 30 to 60 dB gain.

The TVRO receiver gain must be over a wide range. The receiver should be able to process a signal as low as −65 dB (less than 1 microvolt) to −20 dB (100 microvolts). The best answer to Q4 is b, 20 to 50 dB gain. By being able to process very weak signals, a single 10-foot reflector can, with separate horizontal and vertical LNBs (low noise block downconverter), have the LNB outputs split to more than 10 different receivers, as is frequently done at cable TV head ends.

In Q5 the term dithering is used. It is a process used in the transmission to disperse the FM signal over a wide portion of the channel pass band of frequencies. The object is to reduce the likelihood of interference with other terrestrial frequencies in the same band. To accomplish this the transmission is modulated with a triangular low frequency. To prevent the low frequency from causing white levels of the video to pulsate at the triangular modulated low-frequency rate, a dedithering circuit is used to cancel it out in the receiver (Fig. 20-3).

In Q6 the matches are as follows. The dish drive dc voltage is 36 Vdc, a = 6. The LNB supply voltage is 18 Vdc, b = 5. One of the power supply voltages in the receiver is 5 Vdc, c = 3. 5-volt digital pulses are used with differing duty cycles to position the servo motor in the feedhorn; thus, d = 4. 2000 microvolts is 6 dBmv, about what the RF video signal out of the receiver should be; e = 2. +12 volts and −12 volts are most commonly used in TVRO receiver/positioners to allow external coaxial switches to be flipped according to even or odd channel selection (horizontal or vertical LNB selection). The match to f is 1.

In Q7 the terrestrial TV station carrier is not FM modulated. It is AM (amplitude modulated); however, its audio carrier is broadcast using FM modulation. To

20-3 A view of the motor-sensor end of a 24-inch dish drive. Note the white quad-magnets that energize a reed-switch underneath. Cams and limit switches are to the left.

reduce the noise problem associated with the extremely tiny signals used in satellite, FM is used for both the video and audio. The answer to Q7 is a.

The composite out, which is the output used to drive the Videocipher II decoder, contains all of the information modulated onto the TVRO signal. This includes video and audio, the sync portion of the video, the decoder information, as well as any subcarrier audio channels operating on that channel. Question 8 is false.

The filtered video output used on commercial receivers is the video modulation that has the audio for that channel as well as any subcarrier audio removed. It is pure video. Question 9 is false because de-emphasis for audio transmissions means the audio high-frequency information has been boosted electronically at the transmitter. At the receiver a de-emphasis circuit cancels the boost effect to put the audio back in proper balance. By emphasizing and de-emphasizing, the noise pickup is a smaller percent of the transmitted high frequencies. At the receiver de-emphasis reduces the highs, but reduces the interference noise by an equal amount. This leaves the audio back in its original proportions, but the noise reduced.

As pointed out in the Q5 explanation, background flicker can be caused by allowing the dithered video modulation in the TVRO signal to pass through to the video output. The correct response to Q10 is a.

In Q11, drive motors use a lot of power, 2 to 3 amps frequently. This eliminates answer b. LNBs typically use less than $^{1}/_{4}$ amp of dc power. Therefore answer c is not the best answer. The correct response is a.

SCPC stands for Single Channel Per Carrier. A terrestrial AM or FM broadcast station is SCPC, but it is never referred to by that name. Satellite technology uses audio subcarriers for most radio station or audio broadcasts. By using this method, commercial radio stations can be transmitted as subcarriers of the FM video carrier for that channel. Because the TV channel audio is already broadcast as a subcarrier, most often 6.8 MHz, additional audio channels can be used on the same transponder without interference. The location of the audio subcarriers is 5.5 MHz to 8.0 MHz.

SCPC broadcasts do not occupy the same channel as a video broadcast. Many SCPC stations share the 40 MHz spectrum commonly used by one video broadcaster. Each of the SCPC signals is an FM or digital channel itself, not dependent on other carriers or a video carrier of a TV station. TVRO receivers do not process the SCPC signals. Instead a special receiver similar to an ordinary FM receiver (but operating in the satellite frequency band) is used. Question 12 is false (Fig. 20-4).

20-4 A straight-edge across the dish face and a declinometer checking the aiming of the dish.

Aiming the dish reflector is discussed in Q13. It is true that if the polar bar is properly aimed, it will be pointed towards the North Star. If you are installing TVRO in terrain where the north direction is not known and you don't have a compass, and if you don't know the magnetic offset for a compass for the area anyway, locating the North Star might be the best solution to getting the dish mount oriented close enough to begin the aiming procedure.

The answer to Q14 is b. If the orbiting satellites were further from earth, the declination would be less. The Clarke Belt is 22,300 miles from earth. That is about three earth diameters. After adjusting the polar bar to aim toward the North Star (the angle will be about 40 degrees in Indianapolis, Indiana), an additional tilt of about 6 degrees is needed to aim the reflector downward sufficiently to intersect the band of satellites.

Answer b is correct for Q15. Terrestrial Interference filters are needed on about 10 percent of all TVRO installations. Ma Bell and others have microwave relay towers providing telephone communications along routes criss-crossing the country. Unfortunately, the assigned frequencies of the telephone communications are about 10 MHz either side of the assigned C band satellite channel frequencies. Because both communications frequencies have modulated sidebands, each interferes with the other, similar to what happens with adjacent channel interference. This might not have much effect if both satellite and the phone signals were of similar strength. Both start out near 5 watts at their transmitters; however, the signal from the satellites is 22,300 miles away and the phone transmissions are coming from less than 25 miles away. Even when the TVRO dish is not directly in line with the microwave towers, the "monster" phone signals can splatter off utility lines, trees, and metal surfaces, causing severe interference problems for satellite owners.

A method of removing the effects of TI is to narrow the bandwidth of the satellite channel to less than the approximate 20 MHz normally used, by installing a band-pass filter. A 20 MHz video carrier deviating 10 MHz from either side of its assigned carrier frequency would use a spectrum 20 MHz wide, with either end of the spectrum overlapping the phone company carriers (as they are in operation) on either side. Reducing the bandwidth can help reduce the effects of the interference. The bad news is that by reducing the bandwidth the video quality suffers degradation. Sometimes a compromise is best, thus some bandpass filters are tunable. Usually the bandpass filter is made to operate on the 70 MHz second IF stage of the TVRO receiver. By so doing it can act on the IF no matter which of the 24 channels is being tuned to.

Another filter commonly used at the 70 MHz IF is a notch filter. This hi-Q trap deletes a narrow band at the 60 or 80 MHz carrier frequency of the offending phone company transmissions. The notch filter sucks out the offending carrier frequencies, leaving less interference and more of the desired video information.

When all else fails, resonant cavity traps are made that can trap out single frequencies right at the dish. These are more expensive. They work on the waveguide theory where a cavity resonant to an unwanted frequency will effectively short that frequency out.

In Q16, Ku band transponders are not so uniform as the C band transponders. Some formats use only 16 channels on Ku and others use 32 narrower channels. Many receivers are programmed to format the channels correctly once the software is given the name of that satellite. The correct answer to Q16 is a.

Stand-alone Videocipher descramblers have an accessible adjustment pot that allows the unscrambled audio to be set near the same level as that reproduced by the descrambler circuitry. There is no adjustment for the descrambled sound.

20-5 Circuitry of a Uniden 7000 satellite receiver.

The problem in Q18 is tricky because with the current disconnected, the IR drop across the high resistance connection that might be in the coax wire or a defect in the power supply might elude you. With no signal, if the receiver checks OK, most of us automatically change the LNB, thinking it has been zapped. If the LNB has a shorted or open regulator this will correct the problem. If the regulator is shorted, this symptom can be seen by reconnecting the LNB through a two-way signal splitter, then measuring the coax 18-volt supply again. If the regulator is open, no current flows and the 18 volts will remain whether the coax is connected or disconnected. If the coax and power supply are OK, then d is the correct answer for Q18.

Satellite techs in many areas have no UHF stations and thus do not know which TVRO receivers have modulators capable of passing the 400-900 MHz UHF frequencies. Most TVRO receivers do not pass UHF. (The answer to Q19 is b.) This presents a problem in UHF off-air areas. A U/V bandsplitter should be used prior to the TVRO receiver, with the UHF lead going directly to the TV or VCR, and the VHF connected to the TVRO receiver antenna-in RF input to the internal a-b switch. The purpose in connecting the rooftop antenna to the TVRO receiver is so that the TVRO remote control hand unit can be used to switch between rooftop antenna and satellite reception at will.

The last question, Q20, discusses f/d ratio. That is the focal distance to dish depth ratio. The dish depth is measured by putting a string or straightedge across the dish, then measuring the actual depth of the dish. That figure is related to the focal distance, which is a bit further away from the reflector surface. If the dish depth is 2 feet and the focal distance (the distance away from the dish center at which the feedhorn waveguide entry is supposed to be located) is 4 feet, then the ratio is 2 to 1, or 0.5. More common satellite dish f/d ratios are 0.3 and 0.4. The focal point is unknown, unless the dish manufacturer told you. A tech should be able to calculate the focal distance (Fig. 20-6). It takes a difficult-to-remember formula:

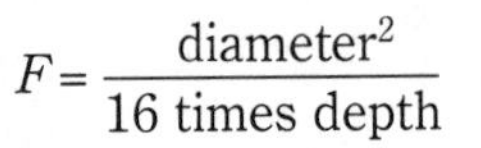

$$F = \frac{\text{diameter}^2}{16 \text{ times depth}}$$

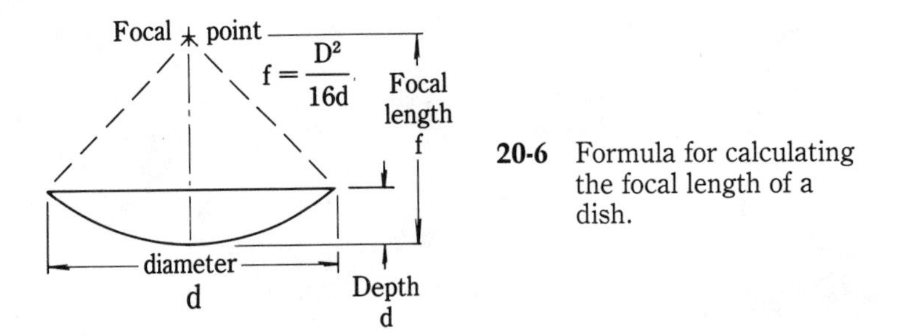

20-6 Formula for calculating the focal length of a dish.

Be sure to use inches, not feet, when working this problem. Like this:

$$F = \frac{\text{diameter (10 inches} = 120 \text{ inches} \times 120 \text{ inches} = 14{,}400)}{16 \times 2 \text{ inches, or } 16 \times 24 \text{ feet} = 384}$$

$$= 37.50 \text{ inches}$$

Do it again:

$$F = \frac{14{,}400}{384}$$

$$= 37.50 \text{ inches}$$

Most satellite technicians also become acquainted with commercial receivers and decoders and other head-end electronics. We have not touched on the differences in Ku band hardware, dual feeds, C/Ku feeds, various actuator and sensor types, or internal receiver circuitry. The complete TVRO CET needs to know about these also. Complete books have been written on these topics.

21 The Telecommunications option

TELECOMMUNICATIONS IS DEFINED AS THE SCIENCE AND TECHNOLOGY OF communication by electrical or electronic means. Telecommunications then should encompass telephone, television, radio communications, avionics, satellite, computer networking, and everything else that electronic communications by voice, data, video, or otherwise does with regard to humans.

In ETA, we have separated communications into more than one category. Satellite is one (the TVRO or Satellite Option), Radio Communications is related to land and mobile two-way, ham, CB, air-to-air, and ship-to-ship voice communications, as well as radar and those other electronics products used by airports and governmental agencies. Telecommunications is the third category. It includes cellular telephone, data communications, fiberoptics (even though FO is used now in just about every part of electronics), microwave phone and datacom (as used regularly by the common carriers), and telephones, which are now more closely related to other electronics products than ever.

Telecommunications does not include cable TV, military radar and fire control communications, avionics, or maritime radio and radar.

Quiz

Q1. The bandwidth of twisted-pair telephone wires will be approximately:

a. 600 Hz
b. 3 kHz
c. 60 kHz
d. 30 MHz

Q2. Half-duplex mobile phones operate on separate transmit and receive frequencies.

a. True
b. False

Q3. Baud rate is:

a. pulses per second.
b. words per second.
c. bits per second.
d. bytes per second.

Q4. Antenna elements are cut shorter than the actual 1/4 or 1/2 wavelength of the desired signal.

a. True
b. False

Q5. A vertically mounted dipole antenna should have a horizontal beam width of:

a. 0 degrees.
b. about 70 degrees.
c. 180 degrees.
d. 360 degrees.

Q6. PBX stands for:

a. public broadcasting exchange.
b. private branch connection.
c. public branch connection.
d. phone billing exchange.

Q7. In determining the beamwidth of an antenna, the points of 50 percent signal strength response are measured.

a. True
b. False

Q8. CTS is a signal abbreviation used in RS 232C nomenclature. It means:

a. clear to send.
b. clear transmitter signal.
c. common telecom system.
d. collect to sender.

Q9. Frequencies used for cellular phones are in the VHF range.

a. True
b. False

Q10. Which of the following can be used for cellular phone communications?

a. 851 megahertz
b. 752 megahertz
c. 292 megahertz

Q11. Match the following impedances:

a. telephone lines
b. TV twin-lead cable
c. CB radio transmission cable
d. cable TV drop line

1. 50 ohms
2. 72 ohms
3. 300 ohms
4. 600 ohms

Q12. A coil in series with an antenna element effectively _______ the electrical length.

a. shortens
b. lengthens

Q13. Ground rods should be located:

a. nearest the antenna mast.
b. nearest the entry point to the building.
c. nearest the utility power service entry point.

Q14. Satellite data communications links use the _______ primarily.

a. C band
b. Ku band

Q15. Signal levels are not part of the RD 232C standards.

a. True
b. False

Q16. Cellular phones transmit with about 8 watts of power.

a. True
b. False

Q17. The IC component in Fig. 21-1 is:

a. an audio amplifier.
b. a voltage regulator.
c. a bridge circuit.

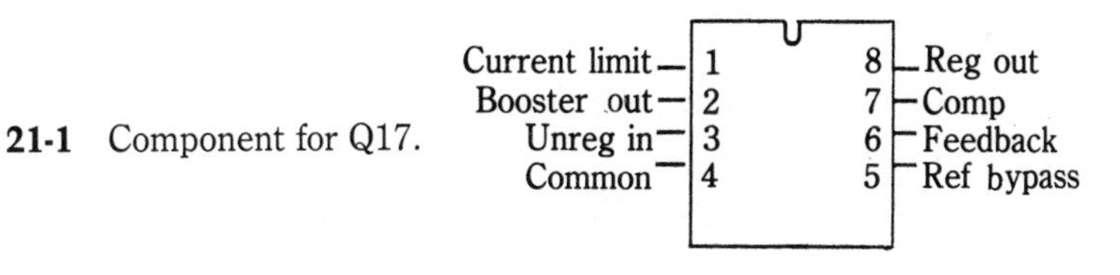

21-1 Component for Q17.

Q18. The circuit in Fig. 21-2 would be used:

a. to regulate voltage.
b. to store data.
c. to amplify audio or video.
d. to drive a display.

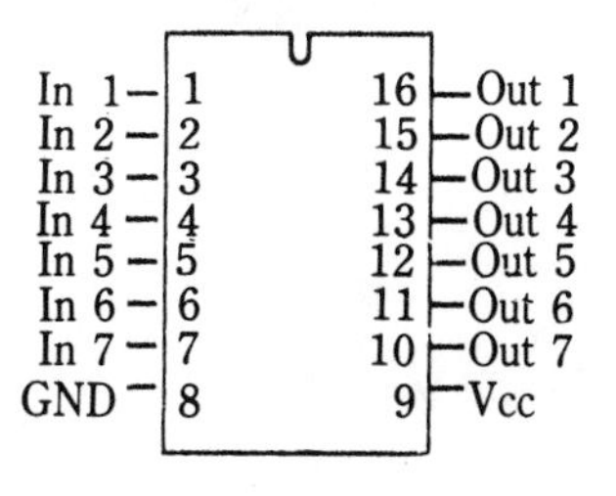

21-2 Drawing for Q18.

Q19. The drawing in Fig. 21-3 represents:

a. a communications interface adapter circuit.
b. a microprocessor.
c. a timer array.
d. a display driver circuit.

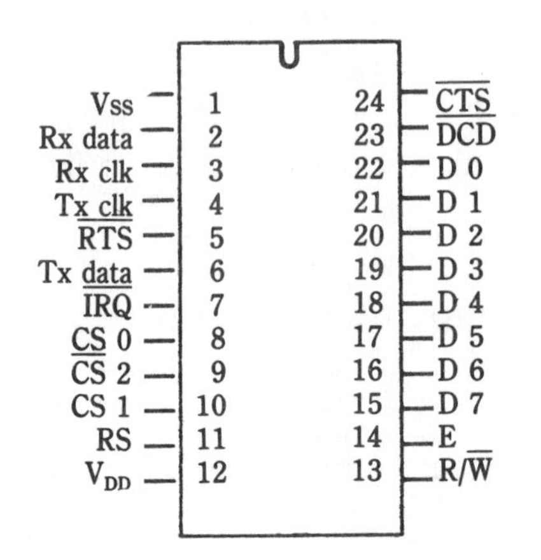

21-3 Drawing for Q19.

Q20. Match the following IC descriptions:

a. 74C76	1. CMOS
b. 74LS107	2. TTL gates
c. 7400	3. Low power Schottky
d. 74S15	4. Super high-speed device

Q21. Audio power amplifier packs might have:

a. 8 in-line connector pins.
b. 10 pins in a DIP configuration.
c. 14 in-line pin connections.
d. Any of the above.

Q22. The package for the transistor or regulator shown in Fig. 21-4 is designated:

a. TO-1
b. TO-3
c. TO-60
d. TO-220

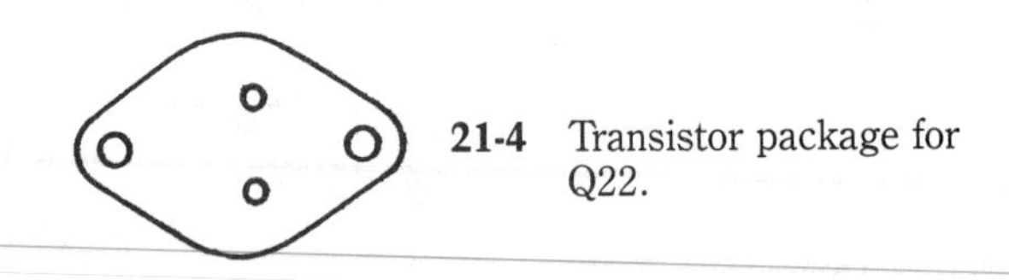

21-4 Transistor package for Q22.

Quiz explanation

The bandwidth needed by phone voice wires is 3 kHz, answer b for Q1.

In half-duplex both transmitters can operate on the same frequency; thus, the answer to Q2 is true.

In Q3, the question is about baud rate. Baud rate is bits per second, answer c. Pulses per second is akin to cycles per second, or hertz. The difference between cycles and hertz is that a count of cycles might not always be cycles per second. Hertz does not mean cycles, but cycles per second.

Antenna elements are shorter than the calculated wavelength by about 5 percent. This is because the velocity of propagation is less in metal conductors than in free space. In free space, the velocity is considered to be the same as the speed of light. Light is an electromagnetic radiation just like radio waves, but at a higher frequency. The answer to Q4 is a.

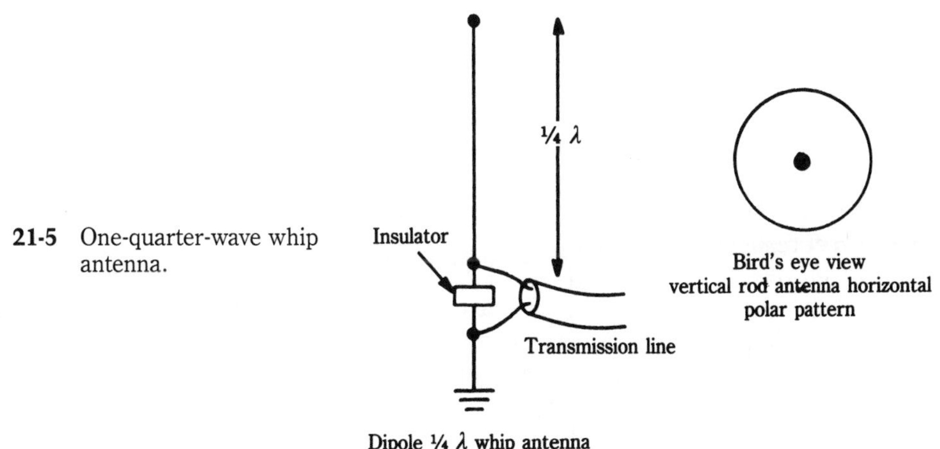

21-5 One-quarter-wave whip antenna.

In Q5 when you speak of a vertically mounted dipole, you visualize a Citizen Band radio antenna or any antenna rod pointed straight up. It is perpendicular to the ground or whatever plane you are working from. As with a CB antenna, the horizontal plane is perpendicular to the vertical antenna rod in all directions, or 360 degrees, a circular plane. The antenna would have no radiation straight up. So with the vertical dipole antenna, the plane of interest is the horizontal. We are trying to communicate with other CB radios located on the same level of earth. Because both antennas are transmitting and receiving parallel to each other, the electromagnetic radiation can be received. It can be received equally in all directions horizontally. Beam width is that portion of the transmitted or received polar pattern on either side of the maximum direction, which does not drop off in voltage response more than 30 percent. This is the half-power level. In this question, beam width is relatively unimportant because it never drops off to the 0.7-voltage level, or half-power level. The power remains constant for the entire 360-degree circle of its polar arc. The correct response to Q5 is b.

PBX stands for private branch connection, answer b to Q6. It is one of those terms nobody bothers to define.

Question 7 addresses beam width again; the correct answer is a. Notice that the 50-percent power level is the same as the 70-percent voltage level.

CTS is one of the easier to remember communications abbreviations. It is used in RS 232C nomenclature and means clear to send (answer a to Q8). While it is common to describe the various computer cable connectors as RS 232, the RS 232 standards as set by industry and governmental groups, contain all the static and dynamic requirements necessary so that one device can be connected and can communicate with another.

Cellular phones operate in a band from 851 MHz to 870 MHz. This is in the 300 – 3000-MHz frequency band, called the UHF band; thus Q9 is false.

In Q10 the only good answer is a, because 851 MHz is in the 851 – 870-MHz band.

Matching in Q11 is as follows:

a = 4
b = 3
c = 1
d = 2

A coil in series with an antenna electrically lengthens the antenna. How else can you get a 9-foot CB antenna down to 3 or 4 feet, a size that autos can use safely. The correct response to Q12 is b.

Ground rods, according to the National Electrical Safety Code, should be 8 feet long and buried at the closest point of entry into the building the receiving equipment is housed in. If there is an equipment building, and the connections are carried from that building outside, then into another building, another ground rod should be installed at the point of closest entry into the second building.

If a copper wire strap is used between the ground block and the ground stake, there should be no bends in the strap because lightning does not like to twist and bend its way to ground.

Ku band, as opposed to C band, is the 500-MHz-wide band of frequencies (11.7 to 12.2 GHz) used mostly for data communications. The best answer to Q14 is b.

Signal levels are a part of the RS 232C standard. Refer to the explanation of Q8.

Cellular phones transmit with 3 watts of power, not eight. The answer to Q16 is b.

The IC component in Fig. 21-1 is a voltage regulator chip. Notice the pins marked Unreg, and reg out? The correct answer to Q17 is b.

In Q18, Fig. 21-2 is a device with 7 input lines and 7 output lines. A ground and a voltage source is shown. While it looks like it might be used to store data (b), there are no control pins; so that is not a good answer. A display driver makes more sense, with a signal-in depending on which LED segment is to be lit, and an output suitable current and voltage wise to correctly energize the segments. The best response for Q18 is d.

Figure 21-3 is not a chip that drives a display, as explained above. It has an R/W or read-write control pin. It isn't a microprocessor because it doesn't have address lines. It isn't a timer array. By process of elimination the answer to Q19 is a, Communications Interface Adapter circuit.

Matching in Q20 is as follows:

a = 1
b = 3
c = 2
d = 4

Question 21 wants to know if you recognize audio output power pack ICs. How many pins do they ordinarily have? They can have any number; 8, 10, and 14 pin in-line packs are common. Therefore the best answer to Q21 is d.

The drawing in Fig. 21-4 is called a TO-3 package. Technicians should be familiar with the common styles for packaging solid-state devices, simply so that they can communicate with other technicians.

22 Test equipment and measurements

THIS CHAPTER CONTAINS QUESTIONS ABOUT MEASUREMENTS AND TEST equipment. Use it as a guide. If you are unsure of the material a particular question relates to and not enough help is contained in this chapter, refer to the reference list at the end of this chapter for further study.

Quiz

Q1. If a 20,000 ohms-per-volt meter movement is to be used to measure 10 volts at full scale, R_x in this circuit (Fig. 22-1) should be:

a. 200 kilohms.
b. 20 kilohms.
c. 9 kilohms.
d. 180 kilohms.

R_x

10 V

20,000 Ω/volt meter movement

V

22-1 Simple voltmeter circuit for Q1.

Q2. The meter in Q1 is used on the 10-volt range to measure the voltage across R1 in the circuit of Fig. 22-2. It will read:

a. 5 volts.
b. 10 volts.
c. less than 0.5 volt.
d. more than 0.5 volt.

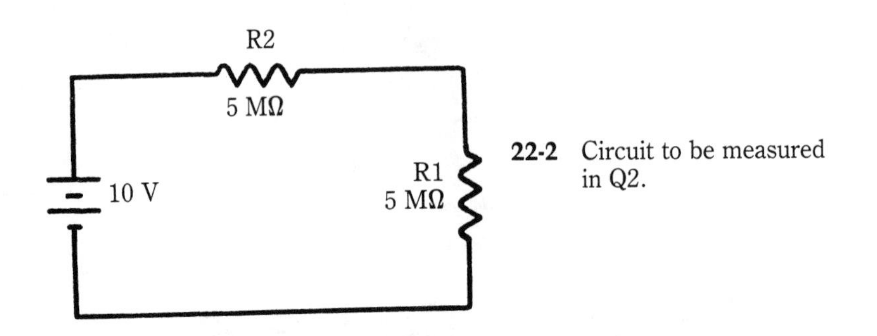

22-2 Circuit to be measured in Q2.

Q3. In the circuit of Fig. 22-3 a low-voltage ohmmeter was used to measure the 2-kΩ resistance. The reading will be reliable:

a. True
b. False

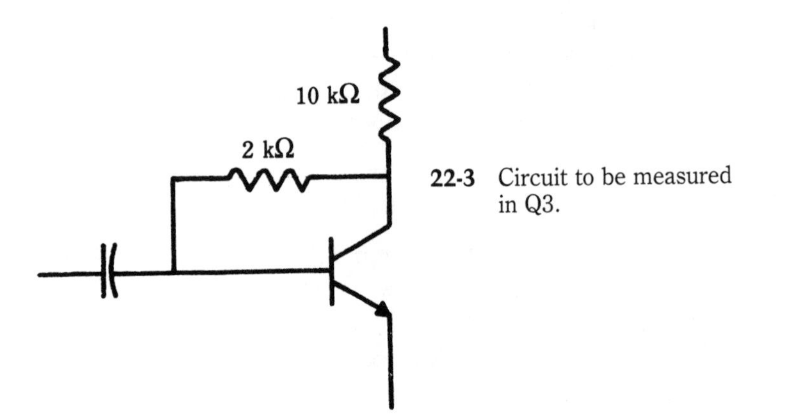

22-3 Circuit to be measured in Q3.

Q4. The 20,000 ohm/volt meter movement of Q1 is going to be used to build an ammeter with 1 milliampere full-scale deflection. A shunt resistance will be needed to bypass most of the current. The shunt resistance should be:

a. 1.053 kilohms.
b. 0.95 kilohm.
c. 20 kilohms.
d. 1 ohm.

Q5. The 1 mA meter of Q4 is to be inserted at X in the circuit of Fig. 22-4. We should expect the measured current to be:

a. less than 0.5 milliampere.
b. more than 0.5 milliampere.
c. not significantly different than 0.5 milliampere.

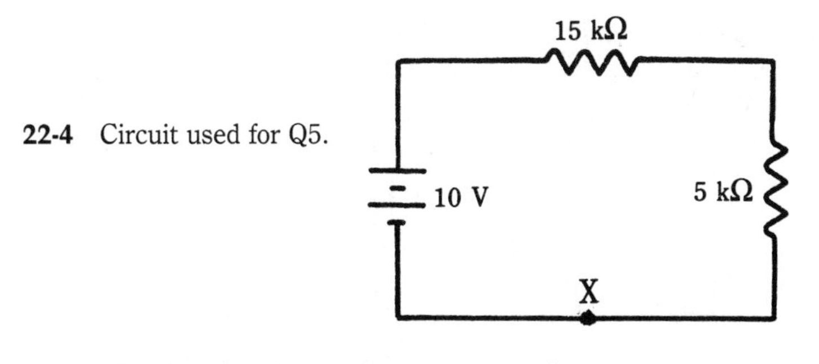

22-4 Circuit used for Q5.

Q6. With an oscilloscope in triggered-sweep operation:

a. the trace starts without regard to the trigger.
b. vertical deflection is triggered by sweep.
c. horizontal deflection is modulated by Z.
d. the sweep voltage starts at trigger.

Q7. Retrace blanking in an oscilloscope is:

a. not used.
b. applied to the Z-axis.
c. used on recurrent sweep only.
d. used to blank vertical retrace.

Q8. Use of a ×10 probe instead of direct connection to a scope:

a. reduces circuit loading.
b. increases the effective input impedance.
c. decreases the signal to the scope.
d. All of the above.

Q9. In the circuit of Fig. 22-5, if you want to see what the waveform at point A looks like along with the scope, you would use:

a. ×1 probe
b. ×10 probe
c. RF probe
d. high-voltage probe

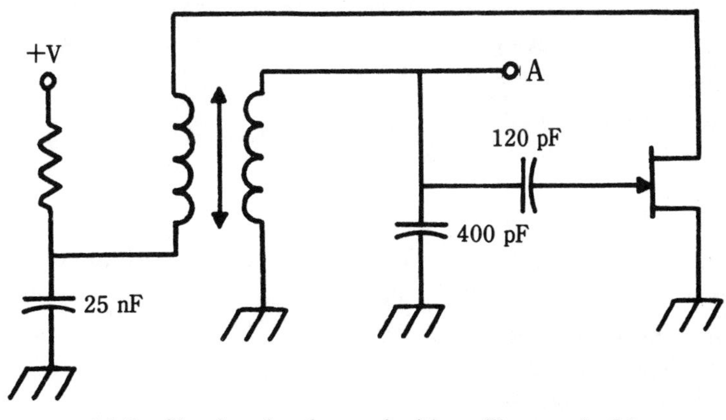

22-5 Circuit to be observed with oscilloscope in Q9.

Q10. If the top of the graticule was set for 0 Vdc, and the sensitivity at 1 V/cm, the waveform shown in Fig. 22-6 when measured with a ×10 probe has a:

a. −2.4 Vdc component
b. −24 Vdc component
c. −0.24 Vdc component
d. 0 Vdc component

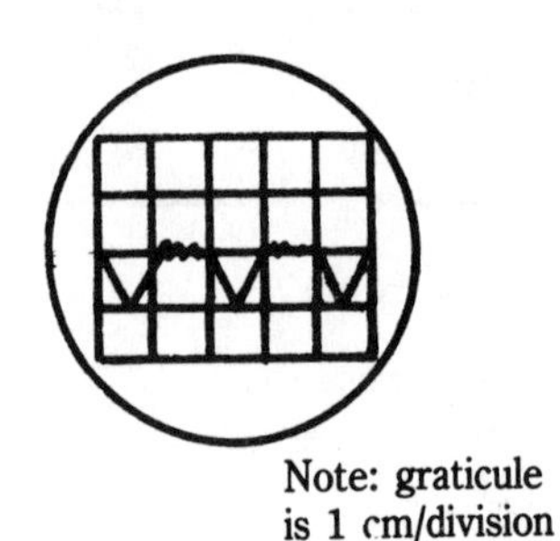

22-6 Waveform measured in Q10.

Q11. In the waveform shown in Fig. 22-6 the peak-to-peak voltage measurement is:

a. 12 V P-P.
b. 1.2 V P-P.
c. 0.12 V P-P.
d. 24 V P-P.

Q12. If the sweep time is calibrated at 10 ms/cm in the waveform of Fig. 22-6 then the frequency of the waveform is:

a. 100 hertz
b. 20 megahertz
c. 50 hertz
d. 10 megahertz

Q13. A digital voltmeter contains a frequency counter. In order to use the counter to measure voltage the input voltage level:

a. is used to scale time.
b. goes to an analog to digital converter.
c. is applied to a galvanometer.
d. Both a and b are correct.

Q14. A function generator delivers sine waves, square waves, triangle waves, and others. If a square wave gets attenuated on all frequencies except its fundamental frequency, the result would be:

a. a sawtooth modified.
b. a "spiked" signal.
c. a sine wave.
d. None of these.

Q15. A circuit is to be aligned using a sweep generator and post marker generator. The marker generator is usually a:

a. continuously sweeping frequency.
b. crystal-generated frequency.
c. constant amplitude signal.
d. None of the above.

Q16. The output of a flip-flop used in a binary counter is being checked with a logic probe. The indicator of normal operation would be:

a. a blinking light.
b. a light that is out.
c. a light that is lit.
d. None of these.

Q17. A logic pulser probe generates:

a. low duty-cycle pulses.
b. logic pulses.
c. various patterns.
d. All of the above.

Q18. A logic current probe is used to:

a. There is no such probe.
b. trace current through wire.
c. check turbidity of water.
d. inject signals into nodes.

Meters

Proper use of meters is always a concern of electronics technicians. This portion of the chapter deals with some aspects of using meters to test electronic circuits. The sensitivity of a voltmeter can be given in two different ways. Each has to do with the amount of current it takes to cause full-scale deflection of the pointer or needle in a D'Arsonval meter movement. Suppose you have a meter with 20,000 ohms/volt sensitivity and the lowest voltage scale is one volt. That is, when the range switch is set at one volt, it takes one volt applied to the meter input terminals to cause full-scale deflection. The current flowing in the meter will be one volt divided by 20,000 ohms or 0.05 milliampere. Suppose there is a 10-volt scale on this meter. If you apply 10 volts and want full-scale deflection, the meter movement must still have only 0.05 milliampere flowing. Hence, you need a series resistance to drop most of the 10 volts to still wind up with one volt on the meter movement to result in 0.05 milliampere through the meter. The series resistance (R_x, in Fig. 22-1) must then be 9 volts divided by 0.05 milliampere, or 180 kilohms.

Meter loading (voltage)

Meter loading is a problem every technician should be aware of. Almost any test instrument will have some effect on the circuit to which it is attached. Good usage minimizes this effect. A common cause of error is current loading by the meter. In Q2, a circuit is to be measured with a VOM. The circuit has rather large value resistors and we should suspect this will be a problem. Without the meter the voltage will divide equally across R1 and R2, but the introduction of the 200-kΩ resistance of the meter in parallel with R1, significantly alters the circuit. R1 of 5 MΩ in parallel with 200 kΩ of the meter is an equivalent resistance of less than 200 k ($\approx$192 kΩ). The new equivalent circuit then looks like Fig. 22-7. The voltage across the 192 kΩ (voltage the meter "sees") is

$$\frac{10\text{ V }(192\text{ k})}{5\text{ meg} + 192\text{ k}} = 0.396\text{ V. Hence, answer c is correct in Q2.}$$

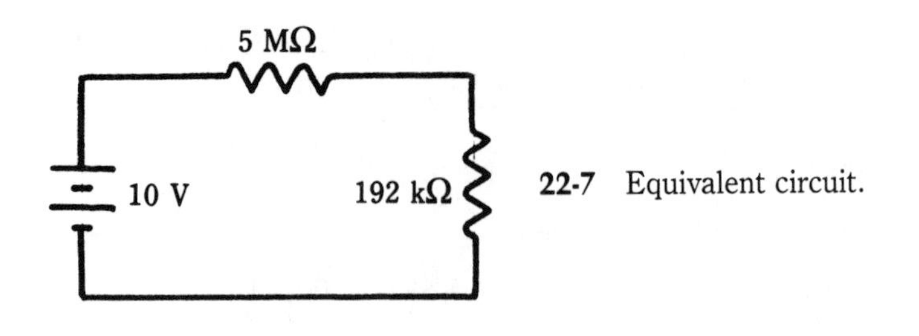

22-7 Equivalent circuit.

Low-voltage ohmmeter

A recent innovation in ohmmeters is the introduction of the low-voltage ohmmeter. This meter is used to measure resistance but uses a low-voltage power source. Remember that ohmmeters (because they operate in unpowered circuits), must provide their own power. Many have two or three batteries for this purpose. An example would be an ohmmeter with a 9-volt battery for high-ohms scales, and a 1.5-volt battery for low-ohms scales; and now ohmmeters with open-circuit output voltages as low as 100 millivolts. The purpose of this low-voltage ohms scale is to measure resistances in parallel with diode or transistor junctions without biasing the junction, thereby giving a parallel current path. The answer to Q3 is therefore a, true.

Current measurement

If a meter movement is to be used in measuring current, a common way to change the range of currents to be measured is to put shunts around the meter. Notice that the meter movement of Q1 (20,000 ohms/volt) could be used directly to measure up to 0.05 mA (or 50 μA). Measuring larger current, however, necessitates a shunt. Question 4 involves calculating the shunt resistance for using the 20,000 ohms/volt meter to measure full scale of 1 mA. A sketch is shown in Fig. 22-8 to help visualize the problem:

$$\text{Because } I_x = 1 \text{ mA} - 0.05 \text{ mA} = 0.95 \text{ mA}$$

$$\text{then } R_S = \frac{1 \text{ V}}{0.95 \text{ mA}} = 1.0526 \text{ k}\Omega$$

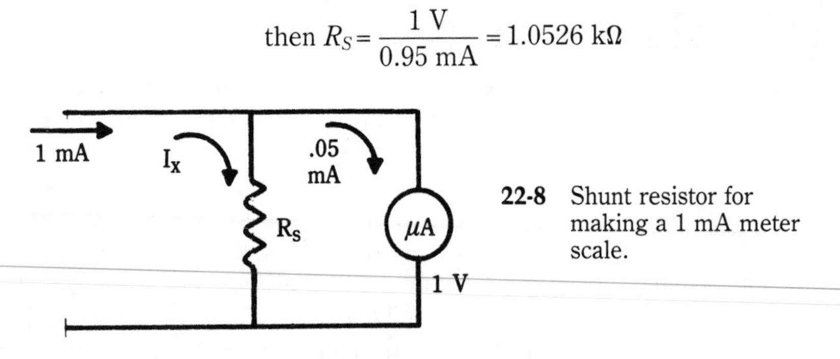

22-8 Shunt resistor for making a 1 mA meter scale.

Loading in current measurement

Using a current meter also causes problems in the circuit being measured. Because the circuit is opened and the meter inserted, the meter's resistance becomes part of the circuit. If the circuit of Q5 is redrawn to include the meter circuit, the result is the circuit shown in Fig. 22-9.

The 10-V battery now "sees" a slightly higher resistance. Instead of 20 kΩ, it sees nearly 21 kΩ; therefore, insertion of the meter reduces the current. If the resistances were accurate, the original current should have been:

$$I = \frac{10\text{ V}}{20\text{ k}\Omega}$$
$$= 0.5\text{ mA}$$

With the increase in the series resistance by insertion of the meter, the current will be approximately

$$I = \frac{10\text{ V}}{20\text{ k}\Omega + 1\text{ k}\Omega}$$
$$= 0.48\text{ mA}$$

Therefore, the answer to Q5 is a. Less than 0.5 mA, or depending on the situation, answer c might be correct; 0.02 mA might not be a significant difference in a particular application. A question with two answers equally correct would *not* be included in the Certification examination. This one was included for discussion only.

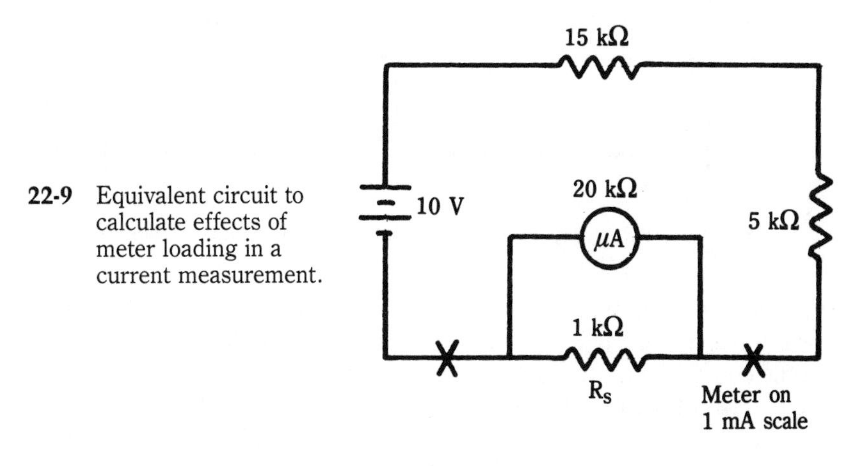

22-9 Equivalent circuit to calculate effects of meter loading in a current measurement.

Oscilloscopes

Because an oscilloscope is an integral part of most electronic troubleshooting, many questions on the Certification exam deal with oscilloscopes. Some questions concern oscilloscope operation and some concern using a scope for measurements. Questions 6 and 7 are examples of operations. In using a triggered sweep scope, you become familiar with various controls that set the generation of a trigger pulse. You might select the trigger source as *External*, *Internal*, or *Line*. On *External*, the trigger generator is connected to the external trigger jack or plug; on *Inter-*

nal, the trigger is generated from the voltage in a vertical channel; on *Line* from the ac power line. Other controls set the *Level*, *Slope*, and *Mode* of operation (such as ac fast, dc, ac slow, etc.). In any case, once a trigger has been generated it in turn starts the horizontal sweep generator which normally moves the electron beam from left to right on the CRT screen or display. Hence, the correct answer in Q6 is d.

A new sweep will not be generated until a new trigger is generated. The time the sweep takes to go from left to right is controlled by the sweep time controls. To illustrate these concepts, examine Fig. 22-10. The trigger point, T, has been set at the indicated point by adjusting for positive slope and appropriate level. Thus, the trigger illustrated in B might be generated.

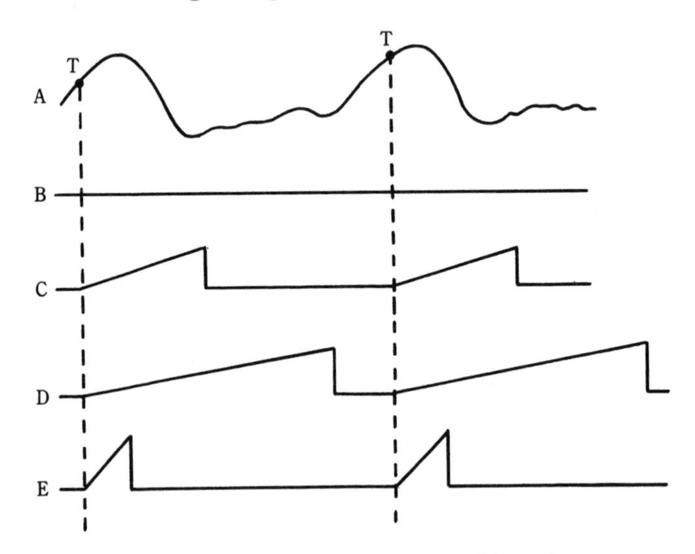

22-10 Waveforms in a triggered sweep scope with various sweep speeds.

Suppose the sweep time is selected to give a waveform shown in C. Notice the waveform starts at the trigger occurrence. Its time duration, however, is set by the sweep time controls. D and E are different sweep-time settings. The displays generated by these different settings are illustrated in Fig. 22-11.

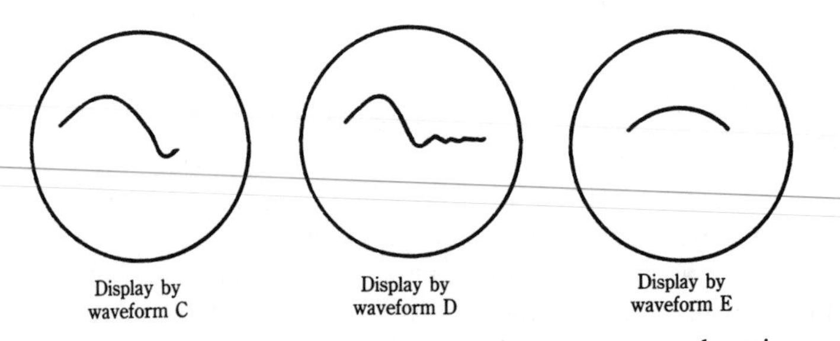

22-11 Waveforms displayed with the various sweep-speed settings.

Z-axis

An oscilloscope has an X-, Y-, and a Z-axis. The vertical deflection is the Y-axis. The horizontal deflection is the X-axis and the Z-axis is the beam intensity. Thus, the correct answer to Q7 is b. The Z-axis is used for blanking horizontal retrace. The Z-axis is sometimes available at the rear of the oscilloscope for external connections.

Probes and loading

When used for measurement, an oscilloscope loads the circuit to which it is attached. A typical scope input impedance is 1 megohm with input capacitance of 30 pF. A ×10 probe for this scope would have an input resistance of about 10 megohms and an input capacitance of 10 pF. A ×100 probe would have input resistance of about 100 megohms and an input capacitance of about 3 pF. The attenuation in a ×10 probe is 1/10. In a ×100 probe it is 1/100. Because a ×10 probe attenuates the signal to the scope, reduces circuit loading of the circuit to be tested, and increases the input impedance, answer d is correct in Q8.

Though the input resistance increases 10 times in a ×10 probe and 100 times in a ×100 probe, the capacitance does not decrease by the same factors. The reason for this is that the cable required to attach the probe to the scope adds capacitance to the input capacitance of the scope. Each length of cable adds capacitance, and the probe itself adds some capacitance. Figure 22-12 illustrates this point.

The probes discussed here have different names to describe the same thing. They are called attenuator probes, low-capacitance probes, and sometimes isolation probes. A high-voltage probe is a special purpose attenuator probe normally used to measure high voltages. Because a high-voltage probe also has low-input capacitance (2 – 3 pF) it can be used where capacitive loading is a problem; such as, the tuned circuits in radios, TVs, communication equipment, etc; however, the large attenuation means that the signal being measured must be pretty large to get a usable scope pattern.

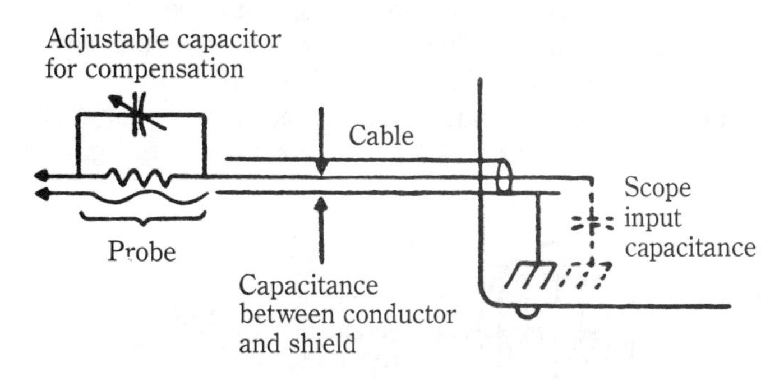

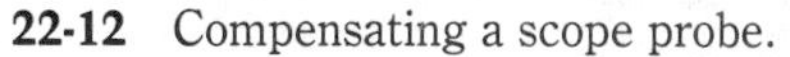
22-12 Compensating a scope probe.

The answer to Q9 should be b (a ×10 probe). However, the addition of about 10 pF to the tank circuit will detune the circuit. If that detuning is a problem, a ×100 probe might be necessary, or even a ×1000 (high-voltage probe).

Special-purpose probes

Other probes are available. One of these is an RF probe (also called demodulator probe or detector probe). This probe contains a detector circuit. RF at the input to the probe is rectified and filtered. The output of the probe is dc or if the input RF was modulated with a low frequency, the probe output is proportional to the modulation. Current probes are also available to measure alternating currents in a wire.

Scope measurements

You will find several questions on the Certification exam dealing with scope waveform and dc measurement. You need to know the effects of an attenuator probe and how to estimate voltages from a scope trace. Questions 10 and 11 are examples. The answer to Q10 is b (−24 V). Here it is necessary to estimate the average value of the waveform and determine how far that average is deflected from the 0 Vdc reference (top line of the graticule in this example). The waveform is reproduced in Fig. 22-13 to clarify this concept. Because the average value is something less than halfway from top to bottom of the waveform (a symmetrical waveform would be exactly halfway between peaks), a line is drawn in at this estimate. The deflection of this average line from 0 is the dc value. If you had the scope in front of you this estimate could be checked by switching to ac coupling. When switched to ac coupling the average value will coincide with the 0 Vdc line. The waveform would move up (in this case) and be above and below the top line. If the estimate is correct the negative peak would be at −6 V.

To more accurately check the dc value, set a convenient portion of the waveform on a graticule line with the vertical position control while using ac coupling. In this case, the negative peaks might be easiest. Then switch to dc (direct coupling). The vertical distance the point moves is the dc component of the waveform. Actually, the deflection is measured in centimeters and the proper scale factors (vertical sensitivity and probe attenuation factor) must be used to determine the voltage. In this case the average value is approximately 2.4 cm down from the top (our dc reference). The sensitivity is 1 V/cm. Therefore, at the input to the scope is a 2.4-Vdc component. The attenuator probe reduces the input signal by 1/10. Therefore, the input voltage has a 24-Vdc component in the negative direction.

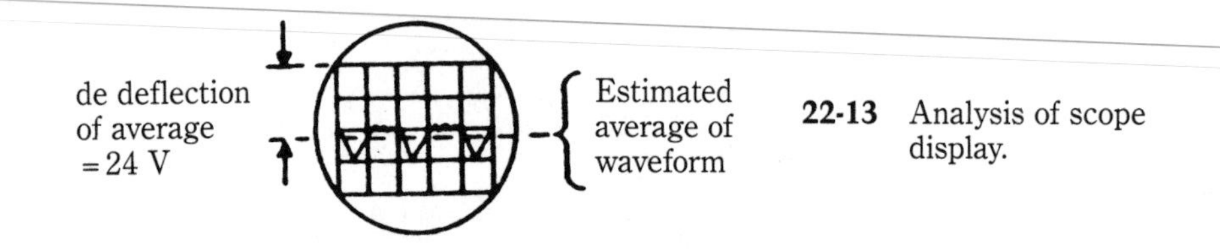

22-13 Analysis of scope display.

All the same sensitivities apply to the ac components of the waveform. In this case, the peak-to-peak deflection is about 1.2 cm. 1.2 cm = 1 V/cm × 10 = 12 V. Question 11 is therefore a, 12 V.

Time and frequency

An oscilloscope is also used to measure time and/or frequency. Triggered-sweep scopes have a very linear sweep in the horizontal direction. That is, the deflection voltage increases at a very constant rate. Therefore time can be accurately measured with an oscilloscope. The sweep-time settings are calibrated and can be determined by reading the scales on the knobs. Some scopes even have a digital readout on the screen that displays the time/cm setting. Once time is measured, frequency can be determined easily. For example, in Q12 the sweep time is set at 10 ms/cm. The period of the waveform is 2 cm long (measured from some convenient point on the waveform to the next occurrence of that point). Because 10 ms/cm × 2 cm = 20 ms, the period of the waveform is 20 ms.

$$f = \frac{1}{T}$$

$$f = \frac{1}{20 \text{ ms}}$$

$f = 50$ Hz, answer c.

Counters

With the advent of inexpensive digital circuits, counters are becoming more and more useful. A wide range of counters are now available. Counters basically operate by counting cycles of an input waveform for a predetermined time period. For example, if cycles are counted for one second, the number of cycles counted is the same as the frequency. It isn't too practical to count higher frequencies for one whole second, i.e., 23 MHz would require a count of 23 million. But if 23 MHz was counted for one millisecond the counter would only count to 23.000. If you only counted for 1 microsecond the counter would only count to 23. By appropriate adjustment of the decimal place or of the multiplier, the 23 would be read as 23 MHz. In a counter with five readout digits, the display probably would be arranged to read "over-range" if attempting to count for one second. It would probably read 23,000 kHz if sampled for a millisecond and 23,000 MHz if sampled for one microsecond. A block diagram for a counter is shown in Fig. 22-14.

The waveform of the input voltage is shaped to provide appropriate trigger pulses for the BCD counter. The counter can count sequentially when a trigger is applied. The timer resets the counter and starts it counting from zero. At a time specified by the front panel settings, the timer will stop the counter and enable the "dump gate" which reads the counter outputs (in parallel) into the BCD to 7-segment decoder/drivers. The decoder drivers in turn light up the appropriate display elements. Many of these circuits are now available on a single IC chip.

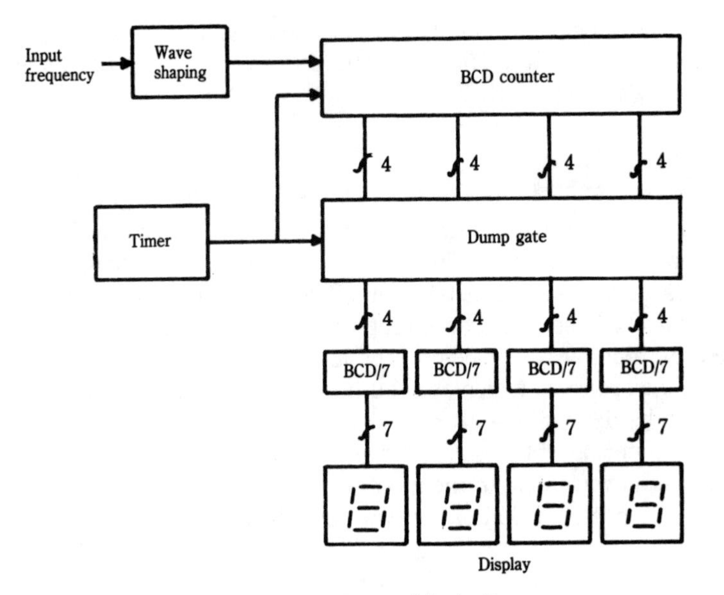

22-14 Frequency counter block diagram.

DMMs

Counters also can be used to construct voltmeters, ohmmeters, milliammeters, capacitance meters, and other special purpose meters. Because of immense production of ICs and improved ICs, these applications are becoming more and more common. The usual functions contained in a digital multimeter (DMM) are: voltmeter, ohmmeter, and milliammeter. The voltmeter is often arranged to measure ac or dc voltages. In all of these applications the desired quantity can be converted to a dc voltage level (by appropriate circuitry) that varies linearly with the unknown quantity. For example, an unknown resistance would be converted to an unknown voltage by applying a voltage to the resistance along with appropriate range resistances in series. The unknown voltage is then applied to a circuit known as an analog-to-digital (A to D) converter. A block diagram of such a circuit is shown in Fig. 22-15.

The circuit functions as follows. The comparator output will allow the clock to trigger the counter until the digital-to-analog (D/A) converter output level matches the input voltage V_x. At that time the counter is stopped and the count transferred to the display. The D/A output increases with each increase of the count in the counter. Shortly after transfer of the count, the counter is reset to zero. This makes the output of the D/A converter go to its lowest level. The comparator switches and the counter starts to count clock pulses. When the D/A output reaches V_x again, the comparator switches and stops the counter. If V_x is larger the count will continue longer. With a smaller V_x the counter will count for less time before the D/A output reaches V_x and shuts off the count thereby transferring the count to the display. Therefore, the transferred count is larger for larger values of V_x and smaller for smaller values.

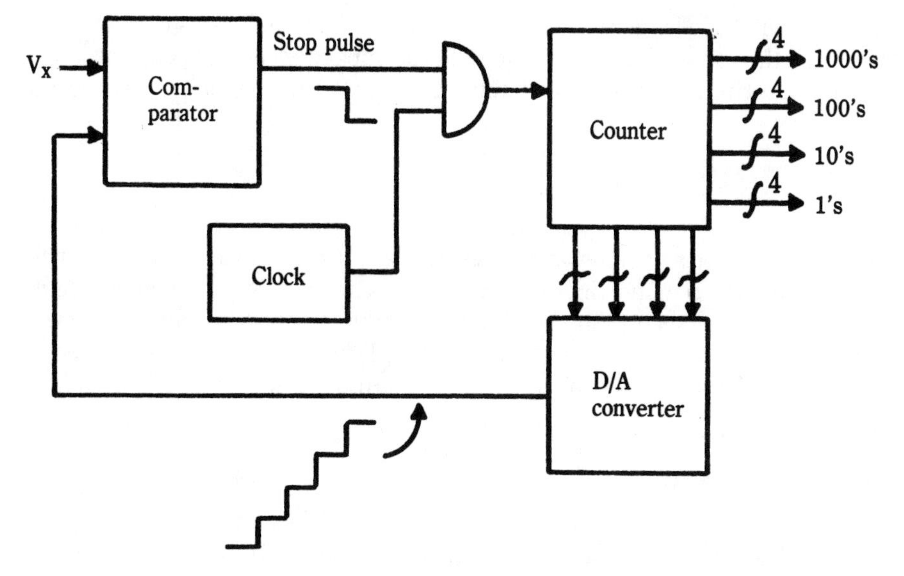

22-15 Digital voltmeter block diagram.

Suppose we are on a 10-volt range and things will be arranged so that the D/A converter will have an output of zero to 1 volt. Resistors and range switches will be arranged so they provide V_x to the comparator from 0 to 1 volt. That is, if 5 volts is applied to the meter, V_x will be 0.5 volt. If 7 volts is applied, V_x will be 0.7 and so on. The D/A output also will range from 0 to 1 volt for the full range of the counter. If the counter can count up to 999, a count of 500 would result in the D/A output of 0.5 volt. A count of 700 would produce 0.7 volt. If your input voltage is 8.5 volts, V_x would be 0.85 volt. When the counter reached 850 the D/A output would be 0.85 volt. The comparator will switch, shutting off the counter and transferring the count to the display, which would be arranged to display 8.50 in all likelihood.

This overall circuitry is called an analog-to-digital converter because an analog signal is changed to a digital count. The most straightforward answer to Q13 is b. A case could be made for answer a being correct, because within the A/D converter, V_x "scales" time by stopping the counter at a time determined by the magnitude of V_x. Therefore, answer d might be the most correct. Again, these sorts of ambiguities will not be part of the Certification exam and are only contained here for discussion purposes. On a Certification exam answer a might be more correct making d the correct answer. Or it might be made completely wrong making b the correct answer.

Capacitance meters

Capacitance meters now can be purchased that use electronic counters. These counters count a frequency for a period of time proportional to the time constant, *T*, of an RC circuit where *C* is the unknown capacitance. If the capacitor is large, the counter will have longer to count and therefore will reach a larger number. The display can, therefore, be calibrated to read the capacitance of the capacitor. *R* can be used in the charging circuit to change the range of the capacitance checker.

Generators

Technicians use all sorts of generators. Sine-wave generators or oscillators are widely used. All sorts of sine-wave generators are used from audio frequencies on into the gigahertz region. Sine-wave generators with a small amount of output power are available (such as grid-dip meters) and larger power generators are also available. Sine-wave generators are used for a large variety of things. They are used to inject signals for signal tracing, for alignment, for calibration, and other purposes.

Function generators

Another class of generators are function generators. These usually come with several types of waveforms available, such as square waves, triangle waves, and sine waves.

Square wave testing of amplifiers and circuits can be an important technique, especially if frequency distortions might be present, like phase-shift and amplitude impairments. For example, if a square wave is applied to a low-pass circuit that allows only the fundamental frequency to pass, the output will be a sine wave at the fundamental frequency. The answer to Q14 is c.

Alignment generators

Many generators are used for alignment of tuned-circuit amplifiers. A generator that generates a single frequency is often used when one frequency is to be amplified by a tuned circuit. For example, the IF circuits in an AM radio can be aligned by applying the IF frequency to the input and adjusting the circuit for maximum passage of the IF frequency. Traps are also often adjusted with application of a single-frequency sine wave signal.

When a tuned amplifier is used to pass a spectrum of frequencies a sweep generator is often used to send signals to the unit under test. The frequencies of interest are swept back and forth usually at a 60-Hz rate. A block diagram is shown in Fig. 22-16.

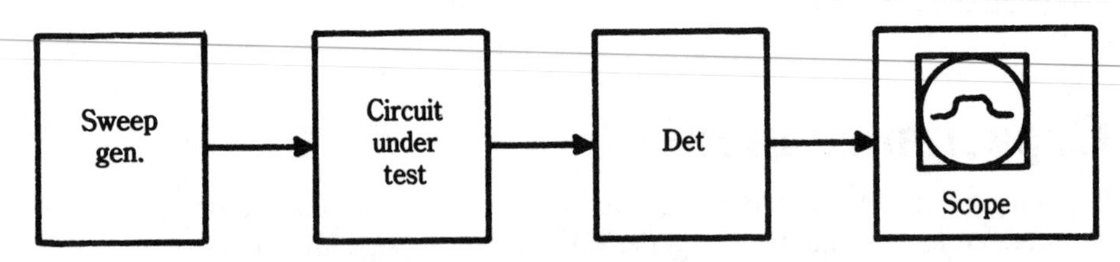

22-16 Sweep alignment generator.

The input amplitude is constant over the range of frequencies. At some of the input frequencies the amplifier under test will be more efficient than others; in other words, at some frequencies more voltage will pass through the circuit under test than will pass through at other frequencies.

If a detector is used at the output of the circuit under test, to separate the variations of amplitude and then send them to an oscilloscope, the circuit's "response" curve can be viewed. Because the sweep generator is changing frequency, it is difficult to construct dials or even counters in the sweep generator to calibrate the frequency being applied to the circuit under test. An alternate scheme uses a separate marker generator (often contained in the same instrument with the sweep generator). The marker generator is usually capable of generating several specific frequencies each controlled by a crystal. Answer b is correct for Q15. Adding the marker generator to the block diagram is shown in Fig. 22-17.

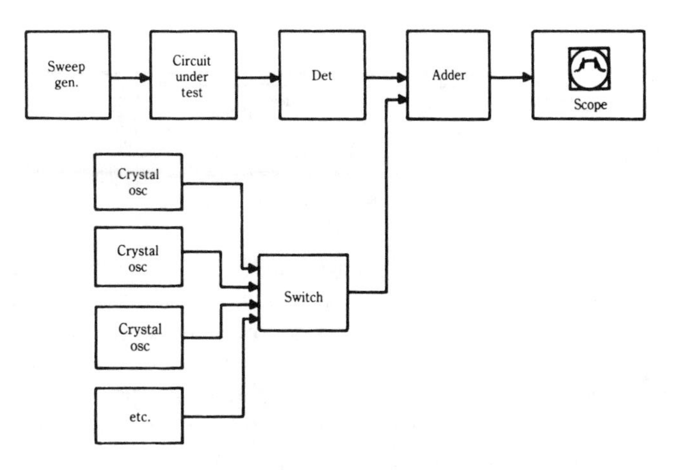

22-17 Post sweep-marker generator.

Logic test equipment

In recent years many new pieces of test equipment have been manufactured to check logic circuits used in all kinds of applications. Logic probes are in this category.

Logic probes

A very simple logic probe could be built, as in Fig. 22-18, with a resistor and an LED.

If the red probe is attached to a positive voltage, the LED will be lit. If the logic level is 0 volts the LED will be out. In a simple probe such as this, make sure that voltages are not excessive for the probe circuit and that polarity is observed.

Many probes are commercially available. One example is the Hewlett-Packard probe shown in Fig. 22-19.

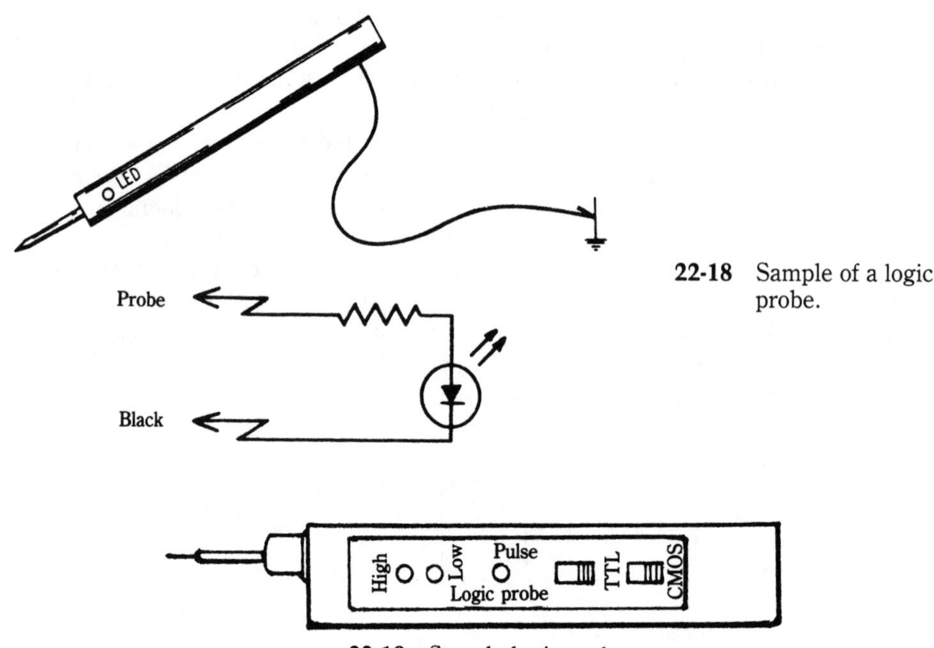

22-18 Sample of a logic probe.

22-19 Sample logic probe.

This probe shows a dim light when attached to an open node, gets bright when attached to a logic "1," and goes out when attached to a logic "0." The logic probe when attached to a node containing pulses (low-, and high-logic levels, sequentially) will blink. The answer to Q16 is therefore a.

A switch position also allows for "memory," that is, a position where a seldom occurring pulse will latch an internal circuit and light the indicator. This is a very useful feature for detecting transients on the power lines and glitches on other lines that should be high or low all the time.

Many companies also make a logic pulser, a kind of portable generator. An example of this kind of probe is shown in Fig. 22-20.

22-20 Sample logic pulser.

This particular probe generates a pulse with a very low duty-cycle. The pulse is very short with respect to the time it is off. This probe can be programmed to generate a variety of pulses, the highest rate of which is 100 pulses per second. At the 100-Hz rate the pulse is on for 10 μs and off 10 ms. A pulse source such as this can be used to cause logic gates to operate, flip-flops to toggle, and other circuits to switch. The answer to Q17 is d. Coupled with the logic level probe, the logic pulser is a powerful troubleshooting tool.

Another probe in the logic probe family is the current probe. This probe costs a little more but is a valuable aid in some troubleshooting situations. A sample of this kind of probe is shown in Fig. 22-21.

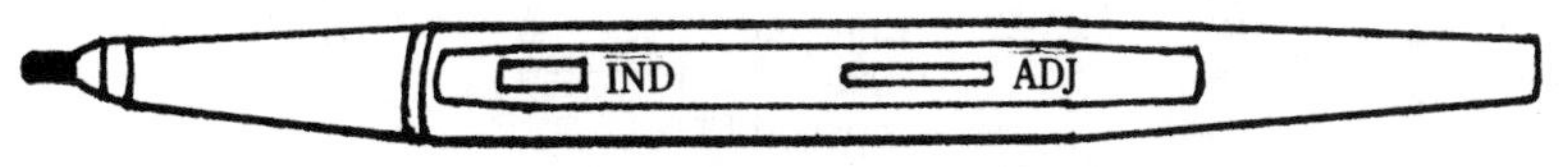

22-21 Sample current probe.

An adjustment on this probe changes the threshold at which the light in the probe blinks. The intensity of the light is proportional to the current flow in a wire adjacent to the probe tip. More current in the wire causes the light to become brighter; less current, less brightness.

On a computer board, one line goes to many different ICs. For example, a data line might go to 32 different RAM memory chips as well as to several logic chips. Suppose one of the chips has a short to ground. How do you find which chip is shorted? Voltage checks will show the line as 0 volts, but any one of the ICs (if shorted) could cause the problem, so could a solder bridge on the line itself. The current probe along with the logic pulser is very valuable in finding this kind of problem. Inject a signal anywhere on the line and trace the current path with the current probe to the shorted spot. Answer b in Q18 is the correct choice.

Logic analyzers

Many logic analyzers are being produced. Because most meaningful information in digital based products is in the voltage levels on many lines simultaneously, it is advantageous to be able to look at several lines all at the same instant. Further, the clarity is extended if all lines can be looked at in steps of time, or sequentially. This is accomplished in logic analyzers by combining the CRT display with digital computer concepts and memory. As patterns occur in the device under test, these patterns are stored in the analyzer for specified times, and then displayed on the CRT in one of several formats; octal, hexadecimal, decimal, binary, timing, or map mode. Analyzers have a possibility of 16 or 32 input lines, depending on the model. The sample display in Fig. 22-22 shows a partial program listing displayed in hexadecimal, with the 16 address lines showing information under the A column and the 8 data lines show information under the B column. The same information is shown in binary format in Fig. 22-23. The timing mode is shown in Fig. 22-24.

By inputting the lower order address lines on 8 of the inputs to the analyzer and causing them to move the CRT electron beam to the right as the magnitude of the address increases; and by inputting the higher order address lines on 8 other inputs to the analyzer and causing these eight to move the beam down as the magnitude of these addresses increases, a memory map is displayed. A picture of such a map is shown in Fig. 22-25 courtesy of Hewlett-Packard.

The map displays the activity of the address bus. If a group of addresses are displayed but they were not programmed this way, give a clue as to how the sequence occurred, leading to a solution to the problem. The analyzer is useful in debugging software problems.

Line No.	A Hex	B Hex
200	06A2	36
001	06A3	00
002	0BF0	00
003	06A4	2C
004	06A5	C3
005	06A6	9B
006	06A7	06
007	069B	34
008	0BF1	03
009	0BF1	04
010	069C	7E
011	0BF1	04
012	069D	FE
013	069E	0A
014	069F	C2

22-22 Analyzer's hexadecimal display (copyright Hewlett-Packard Company).

Line No.	A Bin	B Bin
00	0000011010100010	00110110
001	0000011010100011	00000000
002	0000101111110000	00000000
003	0000011010100100	00101100
004	0000011010100101	11000011
005	0000011010100110	10011011
006	0000011010100111	00000110
007	0000011010011011	00110100
008	0000101111110001	00000011
009	0000101111110001	00000100
010	0000011010011100	01111110
011	0000101111110001	00000100
012	0000011010011101	11111110
013	0000011010011110	00001010
014	0000011010011111	11000010

22-23 Binary program display (copyright Hewlett-Packard Co.).

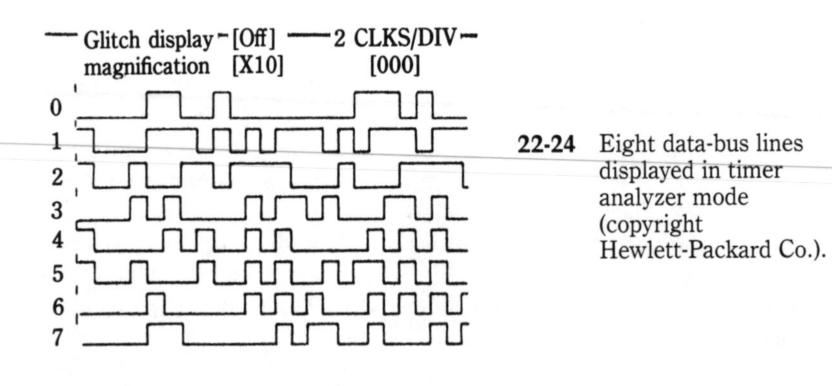

22-24 Eight data-bus lines displayed in timer analyzer mode (copyright Hewlett-Packard Co.).

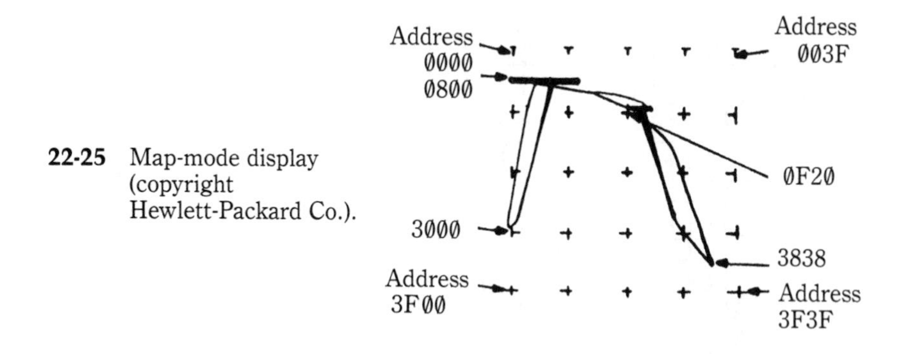

22-25 Map-mode display (copyright Hewlett-Packard Co.).

Spectrum analyzers

Spectrum analyzers are found in some audio shops and in communications work for analyzing output of a transmitter. The purpose of a spectrum analyzer is to display the energy developed by complex waveforms as a function of frequency—the frequency spectrum. In Fig. 22-26 is a graphic presentation of the spectrum caused by modulating a carrier (f_0) with a single frequency (f_1).

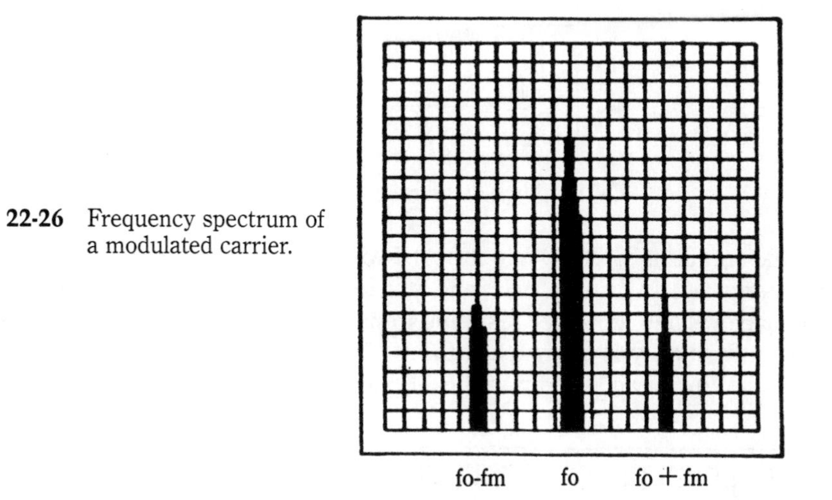

22-26 Frequency spectrum of a modulated carrier.

In testing an audio amplifier, a single frequency is sent to an amplifier. Distortion in the amplifier generates harmonics of this single frequency. A spectrum of this is shown in Fig. 22-27.

A spectrum analyzer uses a section that examines the input waveform and compares it to a sweeping (or frequency-modulated) carrier in a modulator. See Fig. 22-28. As the sweep generator deflects the electron beam in the scope horizontally, it also changes the frequency of the VCO (voltage-controlled oscillator).

As the varying frequency of the VCO is presented to the modulator and compared to the input, sidebands are generated. Suppose the input waveform is a single frequency sine wave at 1 MHz with no distortion. At a particular instant of

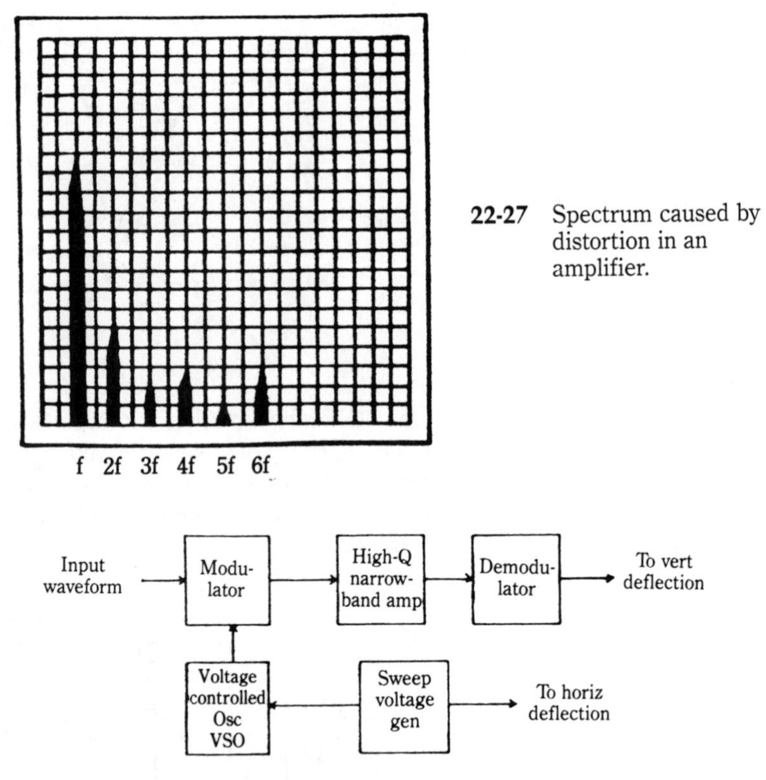

22-27 Spectrum caused by distortion in an amplifier.

22-28 Block diagram of a spectrum analyzer.

time, the output of the VCO might be 0.5 MHz. The modulator will therefore produce energy at 1 MHz, 0.5 MHz, and at 1.5 MHz. These are all outside the range of the narrow-band amplifier resulting in no vertical deflection. As the VCO gets higher in frequency it approaches the 1-MHz frequency of the input. The side-bands start to contain energy within the low-frequency narrow-band (remember

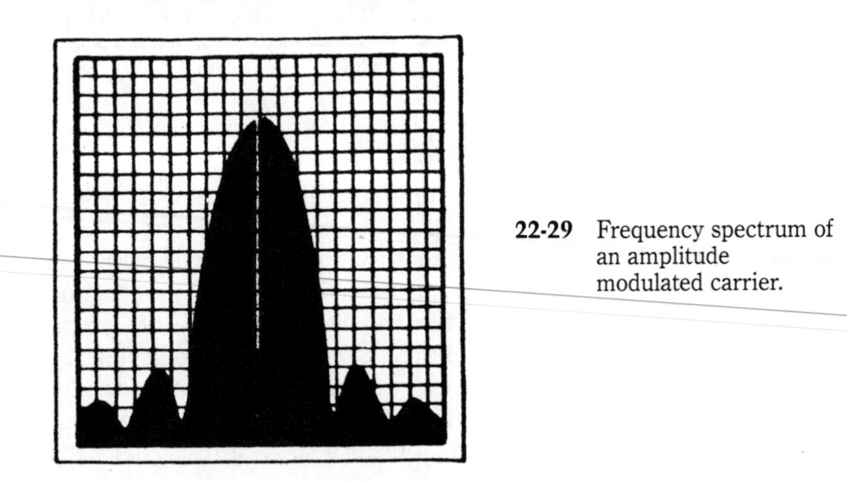

22-29 Frequency spectrum of an amplitude modulated carrier.

the modulator output has sum and difference frequencies) of the narrow-band amplifier. That energy will then cause deflection. The display will show a single spike at the fundamental frequency of the input signal.

A transmitter output contains many frequency components. An amplitude modulated carrier from an AM radio station would have sum and difference frequencies at many frequencies, because the audio modulation contains many frequencies. The bandwidth is limited at the transmitter and regulated by federal standards. A frequency spectrum of an AM modulated carrier is depicted in Fig. 22-29.

Additional reading

Buchsbaum, Walter H.: *Tested Electronics Troubleshooting Methods*, Prentice-Hall, 1974.

Cameron, Derek: *Advanced Oscilloscope Handbook for Technicians and Engineers*, Reston Publishing Co., 1977.

Herrick, Clyde W.: *Electronic Troubleshooting: A Manual for Engineers and Technicians*, Reston Publishing Co., 1974.

Lenk, John D.: *Handbook of Electronic Meters: Theory and Application*, Prentice-Hall, 1969.

Lenk, John D.: *Handbook of Practical Electronic Tests and Measurements*, Prentice-Hall, 1969.

Luetzow, Robert H.: *Interfacing Test Circuits with Single-Board Computers*, TAB Books, 1983.

Prentiss, Stan.: *Complete Book of Oscilloscopes*, TAB Books, 1983.

Robinson, Walter: *Handbook of Electronic Instrumentation, Testing, and Troubleshooting*, Reston Publishing Co., 1974.

Appendix A
ETA-I certification administrators and schools

Michael Buesseler, CET
University of Alaska
3211 Providence Ave.
Anchorage AK 99508
(907) 786-1672

Charles Gilmer CETca
Opelika Tech. College
Opelika, AL 36803-2268
(205) 745-6437

Baby Hall
J F Ingram State Tech.
P. O. Box 209
Deatsville, AL 36022
(205) 826-4325

Patrick A Thomason, CET
Patterson State Tech.
3920 Troy Highway
Montgomery, AL 36116
(205) 288-1080

Michael Lanouette, CET
Southeast College of Tech.
828 Downtowner Loop W.
Mobile, AL 36609
(205) 343-8200

Jerry Cook
Gateway VoTech
P.O. Box 3350
Batesville, AR 72207

Gordon Jobson
Eastern Arizona College
600 Church St.
Thatcher, AZ 85552
(602) 428-1133

Stanley Golowski
1550 N. Stapley Dr. # 52
Mesa, AZ 85202
(602) 644-9531

Robert Derby
DeVry
12801 Crossroads Pkwy. S.
City of Industry, CA 91744
(213) 699-9927

Tom Faro
Victor Valley College
18422 Bear Valley Rd.
Victorville, CA 92392-9699

Javed Rasheed
GTE
2970 Inland Empire Blvd.
Ontario, CA 91764
(714) 945-2325

William Hon
ABC Tech.
76 Shasta St.
Chula Vista, CA 92010
(619) 691-8101

Ron Syth
ITT Tech. Inst.
9700 Goethe Rd.
Sacramento, CA 95827-3262
(916) 366-3900

Kevin J Price, CET
825 E. Hospitality Ln.
San Bernardino, CA 92408
(714) 885-3896

Ron Worley
Mesa College
7520 Mesa College Dr.
San Diego, CA 92111
(619) 560-2638

Romuldo Malarayap
4249 Filhurst Ave.
Baldwin Park, CA 91706
(818) 962-8264

Robert Schwerdtfeger
Cal. State Prison
P.O. Box 8800 CA St. Prison
Corcoran, CA 93212-8309

Michael R. Stevens
ABC Schools
4560 Alvarado Canyon Rd.
San Diego, CA 92120
(619) 280-9933

Mohammad Noury-Khorasani
ITT Tech. Van Nuys
6723 Van Nuys Blvd.
Van Nuys, CA 91405
(818) 989-1177

Fred Bantin, CET
Aims Comm. College
P. O. Box 69
Greeley, CO 80632
(303) 426-7267

Richard Counts
Tech. Trades Inst.
772 Horizon Dr.
Grand Junction, CO 81506
(303) 245-8101

Dick Stevens
DE Tech. Comm. Coll. (Geotown)
318 E. Dupont Ave.
Millsboro, DE 19966
(302) 856-5400

Edward Guary, CET
1110 NE. 4th Ave.
Ft., Lauderdale, FL 33304
(305) 763-2964

Chrys Panayiotou
Florida Advanced Tech. Ctr.
250 Grassland Rd. SE.
Palm Bay, FL 32909
(407) 951-1060

Farris McCallister
Orlando-Voc. Tech. Ctr.
301 Amelia St.
Orlando, FL 32801
(407) 425-2756

Michael P Castaldo, CET
Southern College
116 Marta Rd.
De Bary, FL 32713
(407) 668-0833

Dr. Tom Somers
Augusta Tech. Inst.
3116 Deans Bridge Rd.
Augusta, GA 30906
(404) 796-6900

Charles Shelton
Honolulu Comm. College
874 Dillingham Blvd.
Honolulu, HI 96817

Ronald Brehmer
NITC
Highway 18W
Sheldon, IA 51201

Robert VanElsen, CETma
Nat'l. Ed. Ctr.
1119 5th St.
W. Des Moines, IA 50265
(515) 674-4365

John Arbuckle
DMACC
2006 Ankeny Blvd.
Ankeny, IA 50021
(515) 294-5060

Douglas Watson
Kirwood Comm. College
6301 Kirkwood Blvd. SW
Cedar Rapids, IA 52406
(319) 398-5695

Lon Lasher, CET
ICCC-Telecommunications
330 Ave. M
Ft. Dodge, IA 50501
(515) 576-7201

Paul Jansson, CET
Boise State University
1910 University Dr.
Boise, ID 83725
(208) 888-7377

Glenn Brusse, CETma
ITT
2142 Roanoke Dr.
Boise, ID 83712
(208) 344-8376

Fred Joyner Jr.
Lively Area Voc. Tech. Ctr.
500 N. Appleyard Dr.
Tallahassee, FL 32304
(904) 487-7460

Henry Stephens, CET
UEI
11501 Donna Dr.
Tampa, Fl 33637
(813) 988-0747

Bill Denni
George Stone Center
13221 Lillian Hwy.
Pensacola, Fl 32506
(904) 944-1424

Charles R. Couch, Jr.
1305 NE. 7th Terrace
Gainsville, Fl 32601

Victor L. Bragalone
Elkins Inst. of Jacksonville
3947 Blvd. Ctn. Dr. #6
Jacksonville, Fl 32207
(904) 398-6211

Lamar Thurman
Walker Tech. Inst.
RR. 2 Box 185
Rock Springs, GA 30739
(404) 764-1016

Harrie Buswell
Georgia Tech.
Chemistry Dept.
Atlanta, GA 30332
(404) 894-4017

Roy Chastain, CET
NGTI
197 N.
Clarkesville, GA 30523-0065
(404) 754-7751

Fred Hunt
Albany Tech. Inst.
1207 Palmyra Rd.
Albany, GA 31701
(912) 436-8742

Lincoln S. Thomason, CETma
ITT Tech. Institute
ITT 950 Lusk St., Box 7567
Boise, ID 83707-1567
(208) 344-8376

Dick Bridgeman
DeVry
566 Norman Rd.
Bolingbrook, IL 60439
(708) 953-1300

Brent Steel
Elgin Comm. College
1700 Spartan Dr.
Elgin, IL 60123
(708) 697-1000

Frank Thornton
Capital Area Voc. Center
Springfield, IL 62704
(217) 529-5431

Gene Fayollat
CC Bell Electronics
1001 S. Morrison
Collinsville, IL 62234
(618) 345-8070

Robert Drake
Drake's TV
2320 Evergreen
Bloomington, IN 47401
(812) 332-6453

Jack Warfield
ITT
4919 Coldwater Rd.
Ft. Wayne, IN 46825
(219) 484-4107

Frank Brattain, CETma
Ivy Tech.
2325 Chester Blvd.
Richmond, IN 47374
(317) 966-2656

Ed Carroll
115 E. Sumner
Indianapolis, IN 46227
(317) 783-1963

Paul Yost
Louisville Tech.
2610 St. Joe Rd. W.
Sellersburg, IN 47172
(812) 246-9732

Arval Donovan
Donovans TV
818 DeWolfe St.
Vincennes, IN 47591
(812) 886-4666

Leon Howland, CET
4624 E. 10th St.
Indianapolis, IN 46201
(317) 357-4575

Dick Schultz, CET
ITT Tech. Inst.
3741 S. Meridian St.
Indianapolis, IN 46217

John Byers
Ind. Voc. Tech. College
7377 S. Dixie Bee Rd.
Terre Haute, IN 47802
(812) 299-1121

Earl Brune
P.O. Box 68238
Indianapolis, IN, 46268-0238
(317) 848-9595 (home)

John Meither CETsr
N. KY State Voc. Tech.
1025 Amsterdam Rd.
Covington, KY 41011

Paul Sturgeon
Bardstown Voc. School
126 Creekview Ln.
Rineyville, KY 40162
(502) 348-9096

Bruce Porter
Ayres Inst.
P.O. Box 3941
Shreveport, LA 71103
(318) 868-3000

Barry Thornbury, CET
128 Rustic Manor Av.
Pineville, LA 71360-2634
(318) 640-8302

Steven B. Lumpkin
164 Acadian Dr.
Lafayette, LA 70503
(318) 233-6942

Michael Giovaninni, CET
Springfield Tech. Com. College
31 Brunswick St.
Springfield, MA 01108
(413) 737-6034

Ted Sobocienski, CET
26 Carriage House Path
Ashland, MA 01721
(508) 881-3999

John Ryan
ITT
ITT 1671 Worcester Rd. #100
Framingham, MA 01701

Jesse Leach, Jr. CET
231 N. Hammonds Ferry Rd.
Linthicum, MD 21090
(301) 789-8668

Ron Roane
NEC Temple
3601 O'Donnell St.
Baltimore, MD 21224
(301) 675-6000

Bob Del Raso
Nat'l. Ed. Ctr.
2620 Remico
Wyoming, MI 49509
(616) 538-3170

Paul Mason, CET
Brooks Regional Corr. Fac.
2500 S. Sheridan
Muskegon, MI 49443
(616) 773-9200

Michael Korbakis
32 S. Amber Rd.
Scottsville, MI 43454-3609
(616) 866-9132

Ron Coxon
Nat'l. Ed. of Tech.
2620 Remico SW.
Wyoming, MI 49509
(616) 538-3170

Michael L. Mountain, CET
MoTech Education Center
1211 Sheridan
Plymouth, MI 48170
(313) 459-2913

Lee Hudson, CET
Nat'l. Inst. of Tech.
2620 Remico SW.
Wyoming, MI 49509
(616) 538-3170

John Baldwin
Riverland Tech. College
1225 SW 3rd St.
Faribault, MN 55021
(507) 334-3965

Stanley Blankenship, CETsr
1105 W. Center St.
Oronco, MN 55960

John Slama
Northwestern Elec. Inst.
825 E. 41st Ave. NE
Columbia Heights, MN 55421
(612) 781-4881

Tom Anderson
Willmar Area Voc-Tech.
604 NW 33rd St.
Willmar, MN 56201
(612) 235-5114

Terry Stivers, CET
ITT
13505 Lakefront Dr.
Earth City, MO 63045
(314) 298-7800

Don Balkenbush, CET
Nicholas Career Ctr.
RR 2, Box 43
New Bloomfield, MO 65063

Robert Paynter, CETma
103 Cedar Ave.
St. Louis, MO 63119
(314) 644-9292

Robert Arnett
Holmes Junior College
Goodman, MS 39079
(601) 472-2314

Ted Borgart-Chm Engr.
Southern Station
USM Box 5171
Hattiesburg, MS 39401

Ron Cole
Phillips Jr. College
942 Beach Blvd.
Gulf Port, MS 39507-9905

May School-Blaine Weston, CET
May Schools
1306 Central Av., Box 127
Billings, MT 59102-5531
(406) 248-4888

Gary Novasio
May Schools
Box 127
Billings, MT 59103
(406) 259-7000

Raymond Long
ECPI
7015 Albert Pick Rd.
Greensboro, NC 27409
(919) 665-1400

Carlos Surrat
Surry Community College
Dobson, NC 27017
(919) 386-8121

Steve Burcham
Surry Community College
RR 1, Box 20
Lowgap, NC 27024
(919) 352-4052

Charles Landrum
RR 1, Box 55A
Woodland, NC 27897
(919) 587-9411

James F. Leahy, CET
ECPI
1121 Woodbridge Ctr. Dr. #150
Charlotte, NC 28217
(704) 357-0077

Michael Hamby
ECPI
1121 Woodridge Ctr. Dr. #150
Charlotte, NC 28217
(704) 357-0077

Gordon Koch
Mid Plains Vo. Tech.
I 80 & Hwy. 83
North Platte, NE 69101
(308) 532-8740

George Savage, CETma
Box 39
Doniphan, NE 68832
(402) 845-2298

Peter Vezzosi
DeVry Tech. Inst.
479 Green St.
Woodbridge, NJ 07095
(201) 634-9510

Joseph Dambrauskas
17 Plainfield Ave.
Piscataway, NJ 08854

Henry D. Moher
Cittone Inst.
A4 Dalbert St.
Carteret, NJ 07008
(908) 969-1074

Henry Wallace
Star Tech. Institute
1386 S. Delsea
Vineland, NJ 08360

George West, CETsr
7704 Primrose Dr. NW
Albuquerque, NM 87120

Alfred Rodney, CET
Motorola
1190 Albany Ave.
Brooklyn, NY 11203
(718) 629-2349

Beverly Leatherbarrow
Maryvalle Comm. Ed. Ctr.
777 Maryvalle Dr.
Cheektowwaga, NY 14225
(716) 631-9630

Richard Hedrick
H.B. Ward Occup. Ctr.
970 N. Griffing Ave.
Riverhead, NY 11901

Arlyn Smith, CETsr
Alfred State College
S. Brooklyn Ave.
Wellsville, NY 14895

Charles Bechtold
Tompkins Comm. College
170 North St.
Dryden, NY 13053
(607) 844-8211

Don King
ITT
655 Wick Ave.
Youngstown, OH 44501
(216) 747-5555

Jim Arcaro
Elect SVC Inst.
25301 Euclid Ave.
Euclid, OH 44117
(216) 289-1299

Charles Morrison
Mocom Electronics
P.O. Box 332
Ashville, OH 43103
(614) 983-3338

Joseph Zevcheck
13020 Harold Ave.
Cleveland, OH 44134
(216) 671-5902

Robert Kushner
15430 Sprague Rd.
Middleburg Heights, OH 44130
(216) 826-4276

Marshall J. Piccin
Belmont Tech. College
120 Fox-Shannon Pl.
St. Clairsville, OH 43950

Harold Alexander
Jefferson Co. Joint Vo. School
RR 1, County Hwy 22A
Bloomingdale, OH 43910

Basil Collins
Francis Tuttle Vo. Tech.
12777 N. Rockwell
Oklahoma City, OK 73142
(405) 722-4499

Ralph Greer
OK State Tech. University
6910 Bernadine Ln.
Oklahoma City, OK 73159
(405) 685-8308

Bruce A. Jensen
Portland State University
P.O. Box 751
Portland, OR 97207-0751
(503) 725-3621

Jack B. Fisher
ITT Tech
6035 NE 78th Ct.
Portland, OR 97218
(503) 255-6500

Robert Amell, CETsr
National Ed. Center
5650 Derry St.
Harrisburg, PA 17111
(717) 564-4112

Ronald Lettieri, CETma
433 E. Drinker St.
Dunmore, PA 18512
(717) 894-6353

Paul Ryan
Mercer Co. Area Voc. Tech.
RR 58, Box 152
Mercer, PA 16137

Jim Gilmore
Central Westmoreland Area
240 Arona Rd.
New Stanton, PA 15672

Glen Elliott
Greater Johnstown VoTech.
445 Schoolhouse Rd.
Johnstown, PA 15904-2998
(814) 266-6073

Leei Mao
Greenville Tech. College
602 Del Norte Rd.
Greenville, SC 29165
(830) 250-8000

Clark Adams, CETsr
Lake Area Technical College
409 2nd St. NE
Watertown, SD 57201
(605) 882-2018

Bob Hoffman
SE Vo. Tech. Inst.
212 Pam Rd.
Sioux Falls, SD 57105
(605) 331-7632

Gregory Dowdy, CET
ITT
ITT-441 Donelson Pike
Nashville, TN 37214-3526
(615) 889-8700

Charles Callis
School of Engineering
University of Tenn-Martin
Martin, TN 38348

John Cummings
State Tech. Inst. of Memphis
5983 Macon Cove
Memphis, TN 38134
(901) 377-4170

Tom Witte
Texas Vo. Schools
1921 E. Red River St.
Victoria, TX 77901
(512) 575-4768

Dennis Stanley
MTI
17164 Blackhawk Blvd. #G
Friendswood, TX 77546
(713) 996-8180

R. Parnell Privette
Eastfield College
3737 Motley Drive
Mesquite, TX 75150
(214) 324-7670

D. C. (Snow) Larson
1308 Aldrich
Houston, TX 77055
(713) 686-6545

Robert R. Rivette, Jr.
Hallmark Inst.
9205 Dartbrook Dr.
San Antonio, TX 78240
(512) 575-4768

Curley Savoy, CETsr
Nat'l. Educ. Center
1008 Bedford Ct.
Hurst, TX 76053
(817) 284-5682

Jacinto Gonzales
P.O. Box 40
San Antonio, TX 78291-0040
(512) 227-8217

Dr. Marshall Dean
El Paso Comm. College
9370 G W North
El Paso, TX 79924

John K. Perkins
ITT
920 W. Levoy St.
Murray, UT 84123
(801) 537-5003

Stephen L. B. Moulding
Un. of Ut Health Science Ctr.
Un. of UT 50 N. Medical Dr.
Salt Lake City, UT 84132
(801) 581-2877

Demint F. Walker
ECPI
5132 Andover Ct.
Virginia Beach, VA 23464
(804) 461-6161

Asa Andrews
389-D Jury Lane
Newport News, VA 23602
(804) 864-1512

Donald Crow
7641 Hoemes Run Dr.
Falls Church, VA 22046
(703) 560-1224

James Knight
ECPI
4303 Broad St.
Richmond, VA 23203
(703) 359-3535

John McPherson, CET
P.O. Box 1347
Grafton, VA 23692
(804) 898-5072

Patrick Pope, CETsr
ECPI
4303 W. Broad St.
Richmond, VA 23203
(804) 359-3535

Russell-Reid Offhaus, CETsr
P.O. Box 1116 Ridge Rd. Ext
Chincoteague, VA 23336
(804) 824-2656

Robert Eley, CET
Highline Comm. College
1145 S. 216 St. Apt F203
Des Moines, WA 98198
(206) 878-3710

Alfred Izatt
Grays Harbor College
2315 Queets
Hoquiam, WA 98550

Dean Thompson, CET
3580 NE 95th
Seattle, WA 98115
(509) 522-5429

Barbara Adams
Lower Columbia College
P.O. Box 3010
Longview, WA 98632-0310

Michael O'Conner
Lower Columbia College
P.O. Box 3010
Longview, WA 98632-0310

William D. Burnette, CET
P.O. Box 268
Otis Orchards, WA 99027
(509) 927-0059

Donald Howell, CETsr
Lake Washington Vo-Tech. Int
11605 132nd Ave.
Kirkland, WA 98034
(206) 820-2122

Richard G. Gunderson
Lake Washington VoTech.
11605 132nd Ave. NE
Kirkland, WA 98034
(206) 828-5600

Steven J. Unger
Lake Washington VoTech.
11605 132nd Ave. NE
Kirkland, WA 98034
(206) 828-5600

Michael Smith
Perry Tech. Inst.
2011 W. Washington Ave.
Yakima, WA 98903
(509) 453-0374

Duane Johnson, CET
Bellingham VoTech. Inst.
2117 Mt. Baker Hwy.
Bellingham, WA 98226
(206) 676-6438

David Bernhoft, CET
Gateway Tech. College
1001 S. Main St.
Racine, WI 53403

Larry Mattioli
Fox Valley Tech. College
1001 Violet Lane
Little Chute, WI 54140
(414) 788-3478

Randy R. Ruesser, CET
6317 50th St.
Kenosha, WI 53144
(414) 654-1756

Michael Smith
Perry Tech. Inst.
2011 W. Washington Ave.
Yakima, WA 98903
(509) 453-0374

Duane Johnson, CET
Bellingham VoTech. Inst.
2117 Mt. Baker Hwy.
Bellingham, WA 98226
(206) 676-6438

David Bernhoft, CET
Gateway Tech. College
1001 S. Main St.
Racine, WI 53403

Larry Mattioli
Fox Valley Tech. College
1001 Violet Lane
Little Chute, WI 54140
(414) 788-3478

Randy R. Ruesser, CET
6317 50th St.
Kenosha, WI 53144
(414) 654-1756

Appendix B

Membership Application

THE PROFESSIONAL ELECTRONICS
TECHNICIANS ASSOCIATION

Name ______________________________ Phone__________
(First) (Middle) (Last)

Address ______________ City__________ State______ Zip________

Age________

Present Employer ______________________ Position__________

Years in Position__________

Types of Activity You Are Now Engaged In:

[ ] Industrial
[ ] Education
( ) Instructor ( ) Student
[ ] Military
[ ] Medical
[ ] Consumer Electronic Service
[ ] Sound
[ ] Radio Communications
[ ] Computers/Office & Equipment
[ ] Telecommunications
[ ] MATV/Antennas/Satellites
[ ] Broadcasting
[ ] Engineering
[ ] Musical Instruments
[ ] Distribution
[ ] Other ______________________

____

In addition to regular ETA membership, check the division your work is most closely related to:

[ ] Educators (EEA)
[ ] Certified Technicians (CTD)
[ ] Canadian Division (ETA-C)
[ ] Communication Techs (CD)
[ ] Medical (BMD)
[ ] Industrial (ID)
[ ] Shopowners (SO)

Enjoy these important benefits as a member of ETA

Membership Decals
ETA Member Wall Certificate
Wallet Identification Card
By Laws
Monthly **Association News**
Monthly **Tech Training** Program
Monthly **Management Update**
Area Technical & Business Seminars
Discounts on Tech Publications
Job Placement Assistance — Jobnet
Certification Examinations
Small Business Administration Assistance
PR Brochures
Life & Health **Insurance** Savings
Annual Technician Convention
Help when you need it.

ETA ANNUAL DUES

Institutions (schools - suppliers) ...	$150.00 per year**
Business Owners	37.00
Employee Technician..........	32.00
Electronics Student	22.00*
Foreign (Air-Mail) Add to your category above	20.00

*Name of School You Are Enrolled In:

____

**Institutional membership includes tech membership for 5 instructors or other employees.

Please Sign This Application and return with your fee:

Signed ______________________

Date ______________ Amt. $__________

OTHER INFORMATION:

Please give your CET, CSI, CSS or FCC No. if

Applicable: __________ __________
(Not Required)

Other Registrations or Honors: __________

____

____

Mail To ...

ELECTRONICS TECHNICIANS ASSOCIATION
602 NORTH JACKSON STREET
GREENCASTLE, INDIANA 46135
(317) 653-8262

Appendix C

Membership Application

THE SATELLITE DEALERS
ASSOCIATION – SDA

Owner Name ______________________________ Phone # (   )__________
(last) (first) (middle)

Address ______________________ City______________ State______ Zip__________

Age________ Years of experience in satellite .... _____
Years in business .... _____

Name of local newspaper______________________________________

Other services your firm is involved in:

[ ] electronic service
[ ] electronic sales
[ ] off-air antennas
[ ] MATV / CCTV / CATV
[ ] Vsat
[ ] Hotel/Motel SMATV work
[ ] Broadcasting
[ ] A/V sales and service
[ ] Auto electronics
[ ] Computer sales/service
[ ] Broadcasting
[ ] Education
[ ] Other ____________________

____

Do you or your technicians hold any professional certifications or licenses?

[ ] CSI
[ ] CET
[ ] Space-SBCA
[ ] FCC license
[ ] State license (   )st?
[ ] Other ____________________

____

Enjoy these important benefits as a member of SDA

BENEFITS

- Membership Card – Wall Certificate – Decal
- ByLaws – internal assn. correspondence
- Newsletter
- Tech-Tip repair hints
- Regional technical and business seminars
- **JOBNET** employee procurement program
- **ONE-CALL** long distance phone discounts
- Annual convention and trade show
- Industry-Watch for your protection
- Industry voice to provide dealer representation to government - industry - the public
- Help when you need it

SDA ANNUAL DUES

Individual dealer.............................. $ 75.00 Annually

State or other trade
Association affiliated member ...... $ 50.00 Annually
(ETA - SBCA - NARDA)

Associate (supplier-supporting
member firm) .... $500.00 Annually

Foreign air mail – Add .... $ 20.00 Annually

(If you have been convicted of a felony the SDA cannot accept your membership)

Sign here ______________________________

Date ________________ Amt. $__________

Mail To ...

SDA
602 NORTH JACKSON STREET
GREENCASTLE, INDIANA 46135
(317) 653-8262

Appendix D
Television frequency listing

Channel Number	Frequency Limits of Channel	Center Frequency of Carrier		CATV
SUB CHANNEL				**CATV**
	18MHz			
A		Picture	19.0	T-9
		Sound	23.5	
	24MHz			
	30MHz			
C		Picture	31.0	T-11
		Sound	35.5	
	36MHz			
	42MHz			
E		Picture	43.0	T-13
		Sound	47.5	
	48MHz			
VHF LOW BAND				
	54MHz			
2		Picture	55.25	
		Sound	59.75	
	60MHz			
3		Picture	61.25	
		Sound	65.75	
	66MHz			
4		Picture	67.25	
		Sound	71.75	
	72MHz			
5		Picture	77.25	
		Sound	81.75	
	76MHz			
6		Picture	83.25	
		Sound	87.75	
	82MHz			
	88MHz			
FM BAND				
	88MHz		88.00	
	108MHz		108.00	
MID BAND				
	120MHz			
A		Picture	121.25	
		Sound	125.75	
	126MHz			
B		Picture	127.25	
		Sound	131.75	
	132MHz			
C		Picture	133.25	
		Sound	137.75	
	138MHz			
D		Picture	139.25	
		Sound	143.75	
	144MHz			
E		Picture	145.25	
		Sound	149.75	
	150MHz			
F		Picture	151.25	
		Sound	155.75	
	156MHz			
G		Picture	157.25	
		Sound	161.75	
	162MHz			
H		Picture	163.25	
		Sound	167.75	
	168MHz			
I		Picture	169.25	
		Sound	173.25	
	174MHz			

Channel Number	Frequency Limits of Channel	Center Frequency of Carrier	
VHF HIGH BAND			
	174MHz		
7		Picture	175.25
		Sound	179.75
	180MHz		
8		Picture	181.25
		Sound	185.75
	186MHz		
9		Picture	187.25
		Sound	191.75
	192MHz		
10		Picture	193.25
		Sound	197.75
	198MHz		
11		Picture	199.25
		Sound	203.75
	204MHz		
12		Picture	205.25
		Sound	209.75
	210MHz		
13		Picture	211.25
		Sound	215.75
	216MHz		
SUPER BAND			
	216MHz		
J		Picture	217.25
		Sound	221.75
	222MHz		
K		Picture	223.25
		Sound	227.75
	228MHz		
L		Picture	229.25
		Sound	233.75
	234MHz		
M		Picture	235.25
		Sound	239.75
	240MHz		
N		Picture	241.25
		Sound	245.75
	246MHz		
O		Picture	247.25
		Sound	251.75
	252MHz		
P		Picture	253.25
		Sound	257.75
	258MHz		
Q		Picture	259.25
		Sound	263.75
	264MHz		
R		Picture	265.25
		Sound	269.75
	270MHz		
S		Picture	271.25
		Sound	275.75
	276MHz		
T		Picture	277.25
		Sound	281.75
	276MHz		
U		Picture	283.25
		Sound	287.75
	282MHz		
V		Picture	289.25
		Sound	293.75
	288MHz		
W		Picture	295.25
		Sound	299.75
	294MHz		
	300MHz		

UHF BAND

Channel Number	Frequency Limits of Channel	Center Frequency of Carrier: Picture	Center Frequency of Carrier: Sound
14	470MHz–476MHz	471.25	475.75
15	476MHz–482MHz	477.25	481.75
16	482MHz–488MHz	483.25	487.75
17	488MHz–494MHz	489.25	493.75
18	494MHz–500MHz	495.25	499.75
19	500MHz–506MHz	501.25	505.75
20	506MHz–512MHz	507.25	511.75
21	512MHz–518MHz	513.25	517.75
22	518MHz–524MHz	519.25	523.75
23	524MHz–530MHz	525.25	529.75
24	530MHz–536MHz	531.25	535.75
25	536MHz–542MHz	537.25	541.75
26	542MHz–548MHz	543.25	547.75
27	548MHz–554MHz	549.25	553.75
28	554MHz–560MHz	555.25	559.75
29	560MHz–566MHz	561.25	565.75
30	566MHz–572MHz	567.25	571.75
31	572MHz–578MHz	573.25	577.75
32	578MHz–584MHz	579.25	583.75
33	584MHz–590MHz	585.25	589.75
34	590MHz–596MHz	591.25	595.75
35	596MHz–602MHz	597.25	601.75
36	602MHz–608MHz	603.25	607.75
37	608MHz–614MHz	609.25	613.75
38	614MHz–620MHz	615.25	619.75
39	620MHz–626MHz	621.25	625.75
40	626MHz–632MHz	627.25	631.75
41	632MHz–638MHz	633.25	637.75
42	638MHz–644MHz	639.25	643.75
43	644MHz–650MHz	645.25	649.75
44	650MHz–656MHz	651.25	655.75
45	656MHz–662MHz	657.25	661.75
46	662MHz–668MHz	663.25	667.75
47	668MHz–674MHz	669.25	673.75
48	674MHz–680MHz	675.25	679.75
49	680MHz–686MHz	681.25	685.75
50	686MHz–692MHz	687.25	691.75
51	692MHz–698MHz	693.25	697.75
52	698MHz	699.25	703.75
		705.25	709.75
		711.25	715.75
		717.25	721.75

UHF BAND

Channel Number	Frequency Limits of Channel	Center Frequency of Carrier: Picture	Center Frequency of Carrier: Sound
53	704MHz–710MHz	723.25	727.75
54	710MHz–716MHz	729.25	733.75
55	716MHz–722MHz	735.25	739.75
56	722MHz–722MHz	741.25	745.75
57	722MHz–728MHz	747.25	751.75
58	728MHz–734MHz	753.25	757.75
59	734MHz–740MHz	759.25	763.75
60	740MHz–746MHz	765.25	769.75
61	746MHz–752MHz	771.25	775.75
62	752MHz–758MHz	777.25	781.75
63	758MHz–764MHz	783.25	787.75
64	764MHz–770MHz	789.25	793.75
65	770MHz–776MHz	795.25	799.75
66	776MHz–782MHz	801.25	805.75
67	782MHz–788MHz		
68	788MHz–794MHz		
69	794MHz–800MHz		
	806MHz		

TRANSLATOR FREQ.

Channel Number	Frequency Limits of Channel	Center Frequency of Carrier: Picture	Center Frequency of Carrier: Sound
70	806MHz–812MHz	807.25	811.75
71	812MHz–818MHz	813.25	817.75
72	818MHz–824MHz	819.25	823.75
73	824MHz–830MHz	825.25	829.75
74	830MHz–836MHz	831.25	835.75
75	836MHz–842MHz	837.25	841.75
76	842MHz–848MHz	843.25	847.75
77	848MHz–854MHz	849.25	853.75
78	854MHz–860MHz	855.25	859.75
79	860MHz–866MHz	861.25	865.75
80	866MHz–872MHz	867.25	871.75
81	872MHz–878MHz	873.25	877.75
82	878MHz–884MHz	879.25	883.75
83	884MHz–890MHz	885.25	889.75

Appendix E
CET and AET certificates

Samples of the Journeyman CET wall certificate and wallet card

Certified Electronics Technician

Registration number CMPNY XOX Specialty Computer

Certified by Electronics Technicians Association International

August 20, 1992
date issued

Electronics Technicians Association, Int.
accrediting agency

Be it known by these presents that

Alfred Rodney
has successfully completed the requirements and technical test to be universally recognized for competence, ability, and knowledge as an Electronics Technician.

Nick Glass
ETA President

Electronics Technician
Registration no. CMP NJ010
ETA electronics technicians association international

Wallace Wilson

Computer — February 2,1992
specialty — date issued

Electronic Technicians Association
accrediting agency

Nick Glass
ETA President

Samples of the Associate CET wall certificate and wallet card

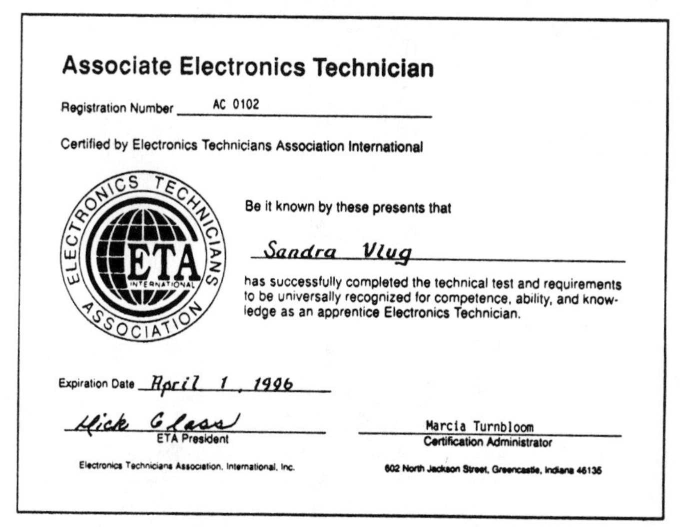

Associate Electronics Technician

Registration Number AC 0102

Certified by Electronics Technicians Association International

ELECTRONICS TECHNICIANS ASSOCIATION ETA INTERNATIONAL

Be it known by these presents that

Sandra Vlug

has successfully completed the technical test and requirements to be universally recognized for competence, ability, and knowledge as an apprentice Electronics Technician.

Expiration Date April 1, 1996

Nick Glass
ETA President

Marcia Turnbloom
Certification Administrator

Electronics Technicians Association, International, Inc.

602 North Jackson Street, Greencastle, Indiana 46135

Associate Electronics Technician
Registration no. AC OXOX
ETA electronics technicians association international

Samone Johnson

9-8-96
expiration date

Electronic Technicians Association
accrediting facility

Nick Glass
ETA President

Appendix F
Master and Senior Electronics Technician certificates

Sample of Master Electronics Technician wall certificate and wallet card

Master Electronics Technician

Be it known that

Electronics Technicians Association

certifies that

Ron Lettieri

on May 14, 1992

successfully completed the requirements

to be universally recognized as a Master Electronics Technician.

SEAL FOUNDED 1978 — ELECTRONICS TECHNICIANS ASSOC., INT'L INDIANA

ELECTRONICS TECHNICIANS ASSOCIATION — ETA

Dick Glass
President of ETA
Leei Mao
Director of Certification
US 1K
Registration number

Master
Electronics Technician
Registration No. US 00
Ron Crow
Bio-Med 1-1-91
Specialty Date Issued
Doug Watson
Certification Administrator
Dick Glass
ETA President

Sample of Senior Electronics Technician wall certificate and wallet card

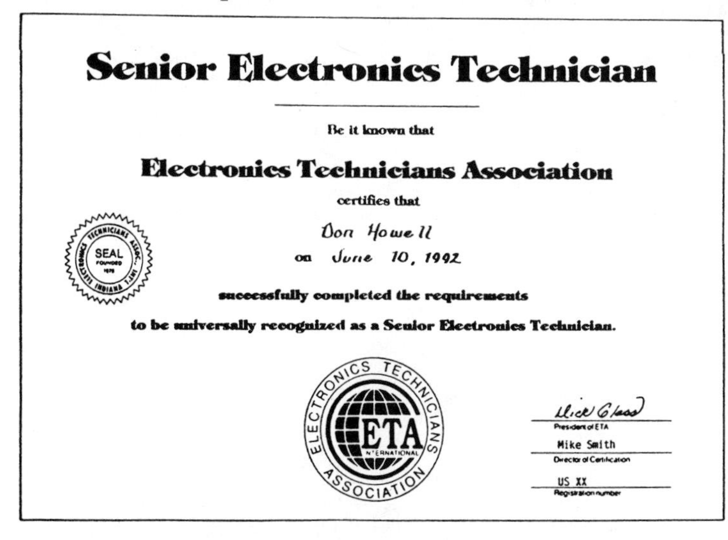

Senior Electronics Technician

Be it known that

Electronics Technicians Association

certifies that

Don Howell

on June 10, 1992

successfully completed the requirements

to be universally recognized as a Senior Electronics Technician.

SEAL FOUNDED 1978 — ELECTRONICS TECHNICIANS ASSOC., INT'L INDIANA

ELECTRONICS TECHNICIANS ASSOCIATION — ETA INTERNATIONAL

Dick Glass
President of ETA
Mike Smith
Director of Certification
US XX
Registration number

Senior
Electronics Technician
Registration No. US 00
Vladimir Klyuchishchev
Senior Electronics Technician 1-1-00
Specialty Date Issued
Ron Roane
Certification Administrator
Dick Glass
ETA President

Appendix G
Certified Satellite Installer

Certified Satellite Installer **CSI**

Registration number CSI IN333

Certified by Electronics Technicians Association International
and the Satellite Dealers Coalition — SDC

CERTIFIED ETA INTERNATIONAL SATELLITE INSTALLER

Be it known by these presents that

Larry Holsey

has successfully completed the requirements and technical test to be universally recognized for competence, ability, and knowledge as a Satellite Technician.

July 16, 1992
Date Issued

Nick Glass
ETA President

SEAL

Roy Chastain, CET
Certification Administrator

Electronics Technicians Association, International, Inc • 602 North Jackson Street Greencastle, Indiana 46135

Certified
Satellite Installer

Registration No. CSI IN777

CERTIFIED ETA SATELLITE INSTALLER

Edward Hiscox

Industrial

Specialty Date Issued

Electronic Technicians Assoc.

Certification Administrator

Nick Glass

ETA President

H
Useful formulas

Resonant frequency formulas

$$F = \frac{1}{2\pi\sqrt{LC}}$$

$$f_{kHz} = \frac{159.2}{\sqrt{LC}}$$

$$L = \frac{1}{4\pi^2 f^2 C}$$

$$L_{\mu H} = \frac{25,330}{f^2 C}$$

$$C = \frac{1}{4\pi^2 f^2 L}$$

$$C_{\mu FD} = \frac{25,330}{f^2 L}$$

Where f is in kHz
L is in microhenries
C is in microfarads

Conversion factors

$\pi = 3.14$ $2\pi = 6.28$
$\pi^2 = 9.87$ $\log\pi = 0.497$

1 meter = 3.28 feet
1 inch = 2.54 centimeters
1 radian = 57.3°

Conversion chart

Powers of two	
1	
2	
4	
8	
16	
32	
64	
128	
256	
512	
1,024	$\cong 1.02 \times 10^{3}$
2,048	$\cong 2.04 \times 10^{3}$
4,096	$\cong 4.09 \times 10^{3}$
8,192	$\cong 8.19 \times 10^{3}$
16,384	$\cong 1.63 \times 10^{4}$
32,768	$\cong 3.27 \times 10^{4}$
65,536	$\cong 6.55 \times 10^{4}$
131,072	$\cong 1.31 \times 10^{5}$
262,144	$\cong 2.62 \times 10^{5}$
524,288	$\cong 5.24 \times 10^{5}$
1,048,576	$\cong 1.04 \times 10^{6}$
2,097,152	$\cong 2.09 \times 10^{6}$
4,194,304	$\cong 4.19 \times 10^{6}$
8,388,608	$\cong 8.38 \times 10^{6}$
16,777,216	$\cong 1.67 \times 10^{7}$
33,554,432	$\cong 3.35 \times 10^{7}$
67,108,864	$\cong 6.71 \times 10^{7}$
134,217,728	$\cong 1.34 \times 10^{8}$
268,435,456	$\cong 2.68 \times 10^{8}$
536,870,912	$\cong 5.36 \times 10^{8}$
1,073,741,824	$\cong 1.07 \times 10^{9}$
2,147,483,648	$\cong 2.14 \times 10^{9}$
4,294,967,296	$\cong 4.29 \times 10^{9}$
8,589,934,592	$\cong 8.58 \times 10^{9}$
17,179,869,184	$\cong 1.71 \times 10^{10}$
34,359,738,368	$\cong 3.43 \times 10^{10}$
68,719,476,736	$\cong 6.87 \times 10^{10}$
137,438,953,472	$\cong 1.37 \times 10^{11}$
274,877,906,944	$\cong 2.74 \times 10^{11}$
549,755,813,888	$\cong 5.49 \times 10^{11}$
1,099,511,627,776	$\cong 1.09 \times 10^{12}$
2,199,023,255,552	$\cong 2.19 \times 10^{12}$
4,398,046,511,104	$\cong 4.39 \times 10^{12}$
8,796,093,022,208	$\cong 8.79 \times 10^{12}$
17,592,186,044,416	$\cong 1.75 \times 10^{13}$
35,184,372,088,832	$\cong 3.51 \times 10^{13}$
70,368,744,177,664	$\cong 7.03 \times 10^{13}$
140,737,488,355,328	$\cong 1.40 \times 10^{14}$
281,474,976,710,656	$\cong 2.81 \times 10^{14}$
562,949,953,421,312	$\cong 5.62 \times 10^{14}$
1,125,899,906,842,624	$\cong 1.12 \times 10^{15}$

Reactance formulas

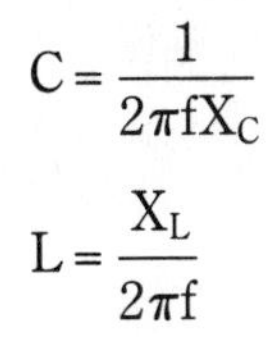

$$C = \frac{1}{2\pi f X_C}$$

$$X_C = \frac{1}{2\pi f C}$$

$$L = \frac{X_L}{2\pi f}$$

$$X_L = 2\pi f L$$

Frequency and wavelength formulas

$$f_{kHz} = \frac{3 \times 10^5}{\lambda_{METERS}}$$

$$\lambda_{METERS} = \frac{3 \times 10^5}{f_{kHz}}$$

$$f_{MHz} = \frac{984}{\lambda_{FEET}}$$

$$\lambda_{FEET} = \frac{984}{f_{MHz}}$$

$0.625\lambda = 225° = 5/8$ wave
$0.5\lambda = 180° =$ Half wave
$0.311\lambda = 112°$
$0.25\lambda = 90° =$ Quarter wave

Resistors in series

$R_{TOTAL} = R_1 + R_2 + R_3 +$

Transformer turns ratios

Primary power = Secondary power

$$\frac{N_P}{N_S} = \frac{E_P}{E_S} = \frac{I_S}{I_P} = \sqrt{\frac{Z_P}{Z_S}}$$

Ohm's law formulas for dc circuits

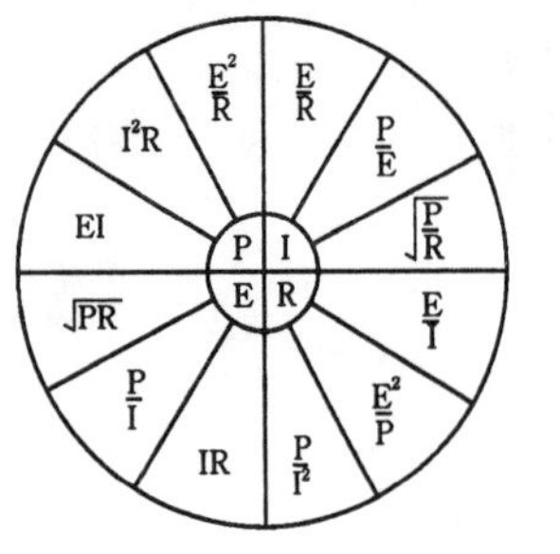

One cycle time duration

10kHz	=	100μsec
20kHz	=	50μsec
100kHz	=	10μsec
200kHz	=	5μsec
250kHz	=	4μsec
1MHz	=	1μsec
4MHz	=	0.25μsec
10MHz	=	0.1μsec

Binary to base 10 conversion

$$
\begin{aligned}
1\,(2^3) &= 8 \\
0\,(2^2) &= 0 \\
1\,(2^1) &= 2 \\
1\,(2^0) &= 1 \\
&+ \\
&\overline{11}
\end{aligned}
$$

Direct power formula

$$P = I^2R$$

Where I is the common point or base current in amperes, and R is the common point or base resistance in ohms

Decibel formulas

Where impedances are equal

$$dB = 10 \log \frac{P_1}{P_2} = 20 \log \frac{E_1}{E_2} = 20 \log \frac{I_1}{I_2}$$

Where impedances are unequal

$$dB = 10 \log \frac{P_1}{P_2} = 20 \log \frac{E_1\sqrt{Z_2}}{E_2\sqrt{Z_1}} = 20 \log \frac{I_1\sqrt{Z_1}}{I_2\sqrt{Z_2}}$$

0dBm (1mW) = 0.774 volts across 600 ohms
0.387 volts across 150 ohms
0.224 volts across 50 ohms

Resistors in parallel

Equal resistors

$$R_{TOTAL} = \frac{R}{n}$$ Where n is the total number of resistors

Unequal resistors

$$R_{TOTAL} = \frac{1}{\frac{1}{R_1} + \frac{1}{R_2} + \frac{1}{R_3} + \ldots}$$

$$R_1 = \frac{R_1R_2}{R_2 - R_1}$$

$$R_{TOTAL} = \frac{R_1R_2}{R_1 + R_2}$$

If the current through a resistor doubles, the power dissipated quadruples

Indirect power formula

$$P = IE(effy)$$

Where I is the final P.A. current in amperes. E is the final P.A. voltage in volts, and effy is the transmitter efficiency expressed in decimal form (70% = 0.79)

Impedance formulas

Series circuits – R & X in series

$$Z = \sqrt{R^2 + (X_L - X_C)^2}$$

Parallel circuits – R & X in parallel

$$Z = \frac{RX}{\sqrt{R^2 + X^2}}$$

Sine wave conversion

Effective value = 0.707 × Peak value
Average value = 0.637 × Peak value
Peak value = 1.414 × Effective value (RMS)
Effective value = 1.11 × Average value
Peak value = 1.57 × Average value
Average value = 0.9 × Effective value (RMS)

Index

T

U

V